TECHNOLOGY AND COSMOGENESIS

TECHNOLOGY AND COSMOGENESIS

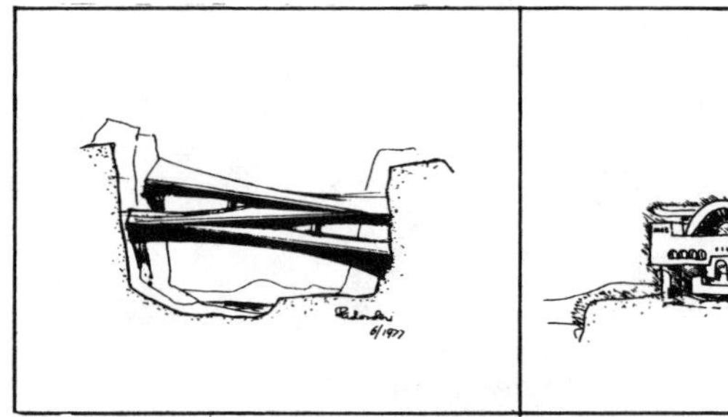

Paolo Soleri

Paragon House Publishers
New York

Published by Paragon House Publishers
2 Hammarskjold Plaza
New York, NY 10017

ISBN: 0-913757-48-9
0-913757-62-4

Library of Congress Cataloging-in-Publication Data

Soleri, Paolo, 1919–
Technology and cosmogenesis.

1. Cosmogony—Miscellanea. 2. Life—Miscellanea.
3. Technology—Miscellanea. 4. Aesthetics—Miscellanea.
I. Title.
BF1999.S572 1985 210 85-19166
ISBN 0-913757-48-9 (pbk.)

Second Printing

Contents

Introduction *vii*
Technology as Cosmogenesis (Process Technology) *1*
A Methodology *15*
Ecology as Theology *18*
Wo-Man—Space—Justice *23*
Art and Environment *33*
Reconciling Esthetics and Theology *39*
A New Kind of Environment *43*
Animism, Technocracy and the Urban Effect *46*
Coevolution and the Seed of the Cosmos *57*
The Energy Dilemma *60*
Organic Structure and the Potential of the New Community *69*
More on the Urban Effect *76*
Dichotomies *79*
Animism and Abortion *84*
Sky Cities *90*
Teilhard and Metamorphosis *97*
Teilhard and the Esthetic *105*
Monasticism and Reverence for Life in the City *112*

Spirit of the Earth *120*
Techno/logy and Theo/logy *131*
Love *140*

Introduction

The assumption of this book is that the mind generates hypotheses. It explores reality and seeks in it some comforting and reassuring handle. Then with the rationales such hypotheses propose the mind proceeds to construct simulation scenarios.

The mind is a categorizing "plant". It tends to characterize its hypotheses as scientific, philosophical, theological, eschatological, esthetic, etc. Often today these distinctions are becoming embarrassing.

The phase of flux and ambiguity of our culture encourages the breakup of niches and the overflow of containers. We model scientific structures and metaphysical ones creep in. We set out on an esthetic terrain and find our minds trapped in a theological mine field.

Astrology, alchemy, animism, shamanism and magic are not just weak flanks to the main flow of today's knowledge. They launch forays into the thick of it. Doubt, confusion, deception, dread, frustration, disillusion ensue. Then why not lean on the revelational certitudes of the Bible, the Koran, the Vedas? "Religion" is on the rise in its most cosmopolitan version that glories in itself as grandest and most tolerant. It says subscribe to God and be aware that this or that God of this or that religion is a limited manifestation of the real, universal divine power. But then it goes on: Do not consider God as an unverifiable hypothesis. Instead embrace and acknowledge this particular manifestation of God as the absolute, indispensable, indisputable truth.

My assumption discards this certitude and this dogmatism. I propose instead that we select one or more of the most promising hypotheses from among the many simulations, regardless of time and place, the grandest included.

Simply put, desirability points to the trust that must eventually generate the true nature of things according to the mode of self-fulfilling prophecy. This type of generation is a process, not a one time revelation. To this we apply the postulate of frugality, that of not dealing with two when dealing with one suffices. Then we can say that the true God is *not yet*, but the true God *will be*, because the creational process is. And we can say that the true God will be true to all the passion for equity, grace and beauty that the mind can muster.

This is a democratic construct on two accounts:

1. It is a notorious grass roots methodology universally applied. God is nowhere to be found. Yet every event that co-authors God produces an extraordinary discipline that eventually prevails. But this discipline is not innate since innateness would then be God in disguise.

2. It implies that whenever an event (particles, organisms, societies, etc.) reaches a higher level of grace (complexity) all other events must eventually find themselves there, at this higher level. In such an escalating chain of events the only ceiling conceivable (and desirable) is the Absolute, and an Absolute which is the proto big bang condition of being converted-transfigured into the last hiatus of time-space, mass-energy. But it is now *all* ablaze in the pure resurrectional logos of Omega, the all inclusive seed and offspring of them all.

There is no creator and creature (the cosmos) in want of it. It is a creature, a reality, that generates from within itself its own transcendence. Because of its youth, most of this creature is still in "darkness". What generates within and issues without begins to assume the responsibility for the creation of Omega, now roughly incarnate in fragments. This Infant God, or Gods (perhaps as many as there are solar systems) are autonomously generating organisms.

They are organisms generating consciousness, consciousness generating anticipations (hypotheses and simulations), anticipations instigating action, action expressing and acting out the manipulative and transformational power of organism. The Infant God is operational.

The degree of pantheism expressed in this Infant God doctrine is tempered if not obliterated by the demonism implied in the behavior of organisms and groups. Violence, hopefully a more and more "reasoned" (rational) violence, underlies transformation (evolution). The Garden of Eden is constantly refuted by the very nature of organism. Organism is the (violent) violation of a pristine status quo for the sake of a contextual status quo. Organism is the mineral complacence forced into the turmoil of the organic. Organism is the relative simplicty of the early organic becoming more and more fragile, yet more and more expectant of the unpredictable. Organism is the relatively tranquil instinctual superseded by the anguish of the mental. The Infant God is restless, demanding, harsh, violent and prodigious. It is a violently permuting, a metamorphozing glory, an inner demon suffering its own genesis into grace. Or is it?

The Infant God doctrine is, if it provides a sequence of events that is synchronic to the evolution of the "mineral reality". It is, if it makes more and more of such mineral metamorphosis into mind at a pace that generates enough complexity, miniaturization, duration, to provide the contracting of the cosmos itself with the "right amount of mind". (I assume the notion that a sufficient amount of matter in the cosmos will be confirmed to warrant the contracting phase of the cosmos itself.) It is, if it thus reaches point Omega where the last hiatus of physical energy suffices to cause the last needed conversion of residual matter into mind, into logos, into verb.

This process of conversion, that turns its origin back to the big bang first instant, puts to shame any other proposed processes of spiritualization. It puts them into the quick fix bottle. The Infant God doctrine cannot accept any residual of unconverted mineral stuff. It cannot accept any second class residual of reality. Can anyone conceive of a "soiled" God with a big greasy spot on an immaculate gossamer mantle? The Infant Doctrine therefore avoids the escape hatch through which inequity might (again) have its own way.

Naturally this doctrine is concerned with a supremely esthetic act since all components become essentially self-justifying and self-explanatory. They are ends in themselves beside being means for the total construct. They are therefore befitting the grace that at the end explodes into the big bang of Spirit. It is an interiorizing big bang generated from the implosion of a reality, the cosmos, and one refuting the endless expansion into an eternal naught.

The metamorphosis of matter into spirit is conducted moment to moment by a techne of transformation. How else could it be? Seen in its totality (a process the Infant God is engaged in), it is a technology of divinization, a theotechnology, the making-creating of that which is not, by that which becomes.

TECHNOLOGY AND COSMOGENESIS

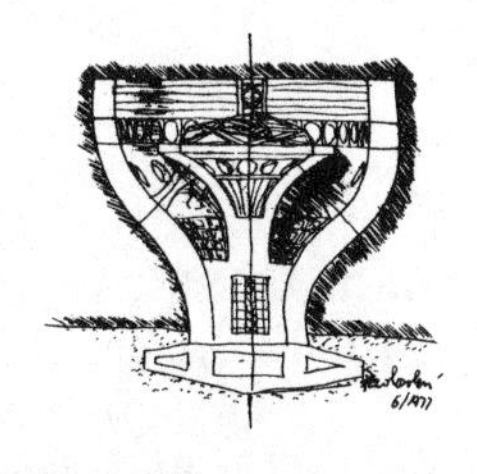

Chapter One:

Technology as Cosmogenesis (Process Technology)

What Is Technology? A Bias

What is technology? This is not an idle question. For we are on the verge of consigning ourselves lock, stock and barrel to technocracy via technology. Our reasons for doing so are unclear. If we frame technology within a "mega machine" of reality we might hit upon some clarifications. This is the ambition of this chapter. Here are two considerations:

First, biological organisms exist and perform through a technology of the flesh. This "bio-technology" is not only immensely subtle but a premonitory of the much younger technology epitomized in *homo faber's* manipulation. A good illustration is the minimotor (electrical?) propelling the cilias of some microorganisms.

Second, we are finding out that millions of years before the advent of the human animal, exo-biological techniques were developed by insects and other organisms, such as, farming by termites and ants, and the use of sticks and other objects by birds and mammals.

In fact, we are better off epistemologically if we forthrightly accept a bio-technology of immense subtlety, present within and "running" all organisms; a latecomer technology. And we humans are its quasi-fiendish exponents and promoters. Then, pursuing this line of reasoning, we begin to order our historical archives and develop a more congruous chronology of things. Somewhat as follows:

First there is the mega-technology of the cosmos. This is the one that classical physics is trying to decipher.

Second is the micro-technology of sub-particles—a wave reality. This is what the new physics is tangled in. It understands its earlier misunderstanding but remains far from understanding.

Third is the micro-technology of "tangible" reality made of things that thermodynamics controls and unwinds (entropy). This is a planet that is half crucible, half crusher of living progression.

Fourth is bio-technology of flesh; of that blue, green and red living stuff that peaks into reflective technology, into the technology of thought.

Fifth is the exo-technology, the flesh-mind that invents its secret follies. This is "technology proper". It is the one that will be assessed in this book.

Sixth is religious technology, that mental construct that attempts to give meaning to the whole (literally) blasted thing (Big Bang). It does so by forward looks concealed in endless backward glances. These are metaphors and myths developed by "technologians".

Seventh is esthetics technology, religion's twin sister. It is a humble undertaking, a work of art that produces nodules of transcendence. The stuff divine is the creation of these technologists of beauty.

Eighth and last are the young latecomers, the technologies of science and philosophy. They are the infants that explain the parental nexus and the cosmic clan. They attempt through torturous paths of shamanism, astrology, alchemy, exorcism, and a postulate of objectivity, to interpret and understand the whole (blasted) thing.

In this chronology, technology is the deus ex machina. Every thing that comes about does so by way of a technology that forges Becoming out of Being. Here technology is time.

One can speak of prototechnology (first, second and third), of cosmic stuff going about in quasi probabilistic ways. But can one speak of post-technology? Perhaps yes but only if one foresees the convergence of biotechnology; with the brain as its fruit. And also if one foresees a supertechnology of the mechanical; with the chemical and the microcircuitry under a spell; under an exorcism of mind-spirit. The ensuing "product" would then be a divine technology of grace-equity-beauty.

This hits at the question beginning to obsess the best contemporary minds. What is the future of flesh, love, grace, sex, beauty, etc.? Can they be transfered into a silicon mind? Or is such a mind above them; a pure, understanding, logical, acting world? Probably this question is poorly posed. For mind is neither flesh nor silicon. It is how flesh speaks to itself and to others. It is how silicon speaks to itself, in its otherness. Perhaps because the self is flesh via mind. While logic is mind via silicone.

Webster defined technology *"as the science of application of knowledge to practical purposes"*. The Delphic oracle could not have been more cryptic: it is science, application, knowledge, practical purpose. One can make them into a thousand windmills. All would spin in its preordained wind.

One interpretation of technology as "the science of application of knowledge to practical purposes" is that of just spelling out a fleshless, knowledgeable, practical, purposeful word or, at least, the legal existence of one.

If such technology is coherent, its existence proposes a "divinization" which is "beyond" sex, grace, love, beauty. It is sexist to the point of disposing of sex. It graces itself in disgrace, it loves to confound love, it is the blind beholder that idles beauty. But isn't perhaps the end "result" of life "the science of application, of knowledge to practical purposes". Cannot we then substitute practical with *real*? Tautology?

Is it the purpose of life to develop technology so as to produce life larger than life? In biological terms *this is just what evolution has been doing for a span of time that has cosmic dimensions* (over 3000 million years). Now with the shift of emphasis from biotechnology to technology, and from evolution to metaevolution (evolution conscious of itself) one has to expect all sorts of crises and disruptions. Ours is an age of transition handling powers never dreamed of before. So much of today's turmoil can be optimistically seen as an excruciating invocation, of the living upon the nonliving in order to make the non-living an "image" of the living. We are brutally dedicated to it and there is no guarantee of success. We may crush life in a technological embrace too robust for the fragile body of a spirit life to be mediated and exorcised out of matter.

Complex Versus Complicated

So how do we handle the demon of technology? I mean what we narrowly recognize as technology.

Why does technology bloody the living stuff of biotechnology so often? Probably because of the mindlessness of technology. I mean this quite literally. The instruments, the equipment we invent, are deprived of mind. They cannot discriminate or do so only coarsely. They are mineral stuff only slightly rearranged and reorganized to perform single (mindless) tasks. They are matter without a master coerced into absurd profiles. To inspect a junkyard is to be confirmed about the coarseness of technology.

This is not technology's sin. It is more the infantilism of a phenomena whose conceiver and maker has a hard time remaining mentally and emotionally together. He is himself only embryonically mature. But the facts might turn out to be that only by way of an excruciating working together of "mind and minerals" will a new level of understanding and creativity become possible.

If the subtlety of the biophysical stuff, life with mind, goes for complexity, it is complex; if the relative subtlety of the mechanical-electrical-chemical stuff goes for the complicated, it is complicated and sometimes to an astounding degree. But it is also only grossly congruent with itself, parts vis-à-vis parts, (a piston and a cylinder for instance) and singularly deaf to most environmental solicitation. (Moods for instance.)

I think that is where the danger of technology lies. It takes over where biotechnology "leaves", and effective as it often is, it is also brutal. It tends often to put things in jeopardy because of a lack of self-control. Leave the stove on and the house might burn down. Leave the centrifuge on and the cream will butter up. Leave the engine on and the train will demolish the station, etc.

But, it is not so much the lack of control (in fact the robot revolution is based on self-control, and self regulation) that might do the flesh in, as it is the technology mode itself (the media tend to be the message). Irresistible as its appeals are, we are willing to submit and to be overtaken. See the automobile mystique. The complex (the living) might then be ultimately victimized by the complicated (the "non-living"), instead of being sustained and incremented by it. That is a regression, a loss of anima. The consequences are visible in the world manifestations of Western society. Without the mediation of mind/purpose, that is the presence of wo-man. the technological device is meaningless. (I adopted the term "*wo-man*" as a small affirmative act. It takes the place of man or mankind.) Some films succeeded for brief moments to convey such a terrifying notion by panning slowly on deserted landscapes of cities and industrial yards.

The manifestations can be thrown into a pot and the pot can be labeled materialism. *Materialism is a pathology of technology*. It is not technology as many bright minds utter confidently. More precisely materialism is the pathology of a society that, unable to discriminate between feasibility and desirability, takes the road of less resistance and drowns life in a sea of per se innocent, re-ordained mineral stuff. It is a shopping center, psyche bugging, collective mesmerizing alchemic tunnel with no visible light at the end.

But de facto technology, as an extension of another technology, the biotechnology of organic life is the effemeralization of reality. It is the process by which a relatively dull universe of mass energy takes time (in space) to re-create, and metamorphosize itself in more subtle, sensitized and ultimately loving occasions.

But if instead a process, via technology, descends the ladder of complexity, the end result might just be the simplification of reality (by simple minded technology). Then the process tends to simplify whatever it touches. See monoagriculture, oil refining, atomic technology. If and when it enters the wo-man condition directly it does the same (I.Q. testing, "social" science, the technology of health, suburbanization, etc).

More and more clearly we are becoming the filters and the retorts (Maxwell's demons of a sort) of matter (the stuff that we can define as the primeval medium) making it into a reorganization of itself, into tools and machines. In that sense technology is the setting up of a neo-nature that we in the west believe to be more direct, more participatory in the venture of becoming (i.e. iron ore becoming useful tools, etc). Materialism, seen in this way as an inclination to an enchanted manipulation of matter, and might well then stand for spirit at the embryonic state; but a state that is more advanced than "determinism", the mineral state of matter still expectant of a fecundation by the mind, via technology.

Our love for the gadget would then be a premonitory condition of our love for the vivification of matter. That this is not sufficient to exorcise the damage of idolatry (making gadgets into idols) is conveyed to us by time; the passing of which is like the steamroller of obsolescence. Its only successful adversaries are the groping of equity and beauty whose categories are capable of resisting obsolescence with the passing of time. Furthermore, technology at its best, is the connective tissue that is more and more indispensable to becoming. Wo-man, as the living unities of body/mind and woman as the aggregation of all things it has invented and developed, are slowly merging into a larger, more complex creature (the urban effect) that is true to the evolutionary inventiveness and the imperative of complexity. There are two dangers: One is the infantility of the body-mind phenomenon (barely three or so million years old). The other is the coarseness, proposed above, of a connective network at its purely embryonic stage.

We are, as individuals, immensely complex bundles of flesh who congregate into complex social, economic cultural systems and institutions. This congregation is then sustained by complicated servo-systems. These are the services and utilities that are expressing themselves and are expressed by a complex physical system, the habitat. In the best of cases this belongs to the esthetic, as also do the arts in general. We are engaged in trying to make the combination of the three manageable, and hopefully self-transcendent (a frightfully tall order).

Feasibility/Desirability

It is somewhat inevitable for machines to acquire their own momentum. (And the feverish minds committed to a balance of terror are ignorant of this fate). We must develop an ethic of consumerism at variance with this contemporary ethic. The cumulative and self-exacting effects of technology's saturation of the wo-man condition, has the mesmerizing and ultimately maiming effect of overthrowing wo-man and enthroning the "idol" machine on the vacated seat. This is not happening in the form depicted by horror movies. It is happening in a contexually ordinary form. It is present and growing within a society at odds with an affluent, buffered, insured, and (by every forecast) happy milieu. A mediocrity of aims automatically produces a mediocrity of means. The worst result is a loss of anima, a mediocrity of spirit, a mediocrity of meaning (well within a mediocre ideology). It is natural for a soul to slowly lose its own dignity and self-respect and to turn to those physically tangible, reassuring and labor saving mechanisms in order to prop up its weakening structure of reverence and self-responsibility. The result is the opposite of the ephemeralization of healthy technology. It is to bury the dance of life in a junkyard of obsolete objects. It is to rob wo-man of resourcefulness and true creativity.

Since manipulation of matter is a love affair with "mother" nature, mainly a male addition *homo faber* we must become more discriminating and direct our passion to the desirable. That means that our exploration (and colonizing) that constantly seeks new limits of the feasible must be stringently disciplined. We must break an irresistible propensity that, through salesmanship, translates feasibility into marketability. Unfortunately, we are learning in most costly ways that what is feasible is usually not desirable. We are now in the process of selling ourselves on the desirability of atomic warfare. Here the death wish of consumerism brings itself face to face with its own nemesis.

The marketplace, through coarse or subtle manipulation, can rapidly make the wo-man environment into a theatre of the absurd. The shopping center is the warehouse where human equity is flushed from wo-man conscience by way of slow, minute, deceptive persuasions pinned, like hidden price tags, on every gadget or thing displayed.

We are truly involved in a monumentally ambiguous project. Optimistically we are engaging the living in processing the non-living, and by so doing upgrading itself. Pessimistically, we are participating in a process that is smothering itself (the living) by an onslaught of matter extravagantly fashioned into pseudo-organizations that clutter the shrinking human landscape. Obscurantism is incipient even in our super-lighted, electrified habitats. Obscurantism is a dissipative condition. But it is not a dissipative process according to Ilya Prigogine's grammar. It is entropy-prone, without offering a more sophisticated redemptive product of mind-spirit. It lacks the discriminatory skill to seek "light", premonition of grace, and esthetic affirmation.

It then should be plausible for discrimination to become of the essence. Discrimination is educated choice (the opposite of segregation). Education/learning would then be a priority, but it cannot be technocratic education. For this education that we are now developing causes illness rather than restors health. A society that puts a premium on an education that involves sensitization rather than information, might be in a better position to descriminate the desirable from the feasible or marketable. In fact, if they turn out to be desirable, it might go so far as to attempt the impossible or improbable.

Stewardship-Transformation

Webster is more helpful in a definition of *techne* than in a definition of technology. *Techne*: "The principles or methods employed in making something or attaining an objective". This fits hand in glove with my proposition that technology is the principle or method(ology) employed in making the genetic structure of reality, and attaining the objective of equity and beauty (divinity). The making (the technology) of such genetic structure, the cosmo-gene-sis, will take all the time needed for the total metamorphosis of reality, and all the space (expanding and contracting), necessary for such metamorphosis to implement, itself; and all the mass-energy available to structure and fuel the process. The fully concluded metamorphosis will imply the termination of time (change), the implosion of space into zero space, and the exhaustion of physical, mass-energy reality. This genesis involving the whole of reality, is a cosmogenesis. It could produce the genes of the cosmos. A gene of the cosmos will be a spaceless, timeless, mass-energyless reality comprehending all the evolution that stands behind it. It will be a resurrection of it all, an Omega Seed of cosmo-gene-sis. Once our question is put in ontological terms, our assessment of technology becomes critical. If wo-man is a phenomenon of relevance within the cosmos, are we to be stewards of the earth or are we to transform the earth in the direction of the improbable?

Should we be modest, tranquil upkeepers of a marvelous ecosystem that is slowly, irresistibly recycling itself? Should we be troublemakers constantly at odds with it; yet impotent in the face of it and scornful of its complex ways? Or are we developing the knowledge and wisdom that will eventually "carry" the ecosystem of the cosmos into spirit? I think the answer will only be found on spiritual and religious grounds, even at the risk of reaching non-religious conclusions.

The adoption by religions of a model of things that is animistic fundamentally precludes the reality of time. A universe vibrant with spirit (from its very inception) that is, a god-created reality can at most propose a circular run, a recycling of matter into that from which it comes. One can stop the recycling anywhere, the cycle is still intact and indestructible. That proposes or admits the reversibility of time. Reversibility is de facto a reality in which consciousness and good and evil are predefined, predestined, fatally positioned and ghostly, clamoring for freedom and transcendence. This is cosmic deception. This is a furious deception with a frozen "perfection", and never the vibrance of matter metamorphosizing into spirit.

This is the basis of the stewardship-maintenance mode. Here every present point on life's cycle is as good as any other. The present is identified by each generation as the "known". It is the loved one, the tangible one, the proven one, the safe one, and the one to be perpetuated. Thus one sees the intrinsic contradicition of this infinite number of "presents" that make up the history of the cosmos Wherein each clamors for its originality and its privileged nature as the rightful end of the historical process. Technology in this mode is not a blessing but a menace. The masses believe they need this mode of technology to produce what is sufficient for their support. This is given further demonstration in the evils of overpopulation and the increased numbers that each new present "arrogantly" brings about. "Things" should remain as they were even though within the cyclical model the *were*, the *is*, the *will be*, are only figments of the imagination that the mind of wo-man is entangled in.

Along with, if not above, such a fatal model one can place a very different one. What if time is real? What if becoming is a creational process? What if the universe has begun to take hold of itself via intellections? What if congruent to the food chain mode of evolution the cosmos is beginning to seek its own genesis? What if it is seeking to create its own seed for instance? What if there was a real beginning and therefore there will be real end? What if becoming is the needed interval for an ignorant media of the beginning in order to become a pure, radiant, loving, beautiful, knowledgeable message, and Being of the end?

If this is so, then technology is the *deus ex machina* of genesis. From cosmogenesis (from the immense mass energy reservoir of the beginning) one would go by way of biogenesis to the infinite grace of the end. Self-effacement would then become a mandatory parameter of technology and a metamorphosis of the "refinement" of matter into spirit. It is then miniaturization. It is the other side of the model of complexification and conjures up with it the metamorphosis of time into duration, because it is only with a progressive effacement (consummation) of time/space that spirit will progressively pervade the real and bring about the "inconceivable" point/moment of pure mind/spirit.

Thus at the limit of this eschatology one sees the cosmos consume into *logos* and the extinctions (consumption) of mass-energy space time. With the cessation of time the miracle of resurrection is actualized. The past, which is the whole of evolution, is incarnated in the ultimate semen. *The technology of transforming cosmos into logos is beyond imagining. But perhaps no other routes are available for consciousness to achieve equity and grace.* The legitimacy of technology is then absolute and being so is terrifying. It must give consciousness the wherewithal to make matter into spirit, into a time-spaceless and mass-energyless reality. After all a dimensional cosmos is a segregative cosmos. A segregative cosmos is an unconscious cosmos. An unconscious cosmos is by definition ignorant of its own fundamental dilemmas.

Summary and Anticipations

If it can be said that the process—the becoming—the beat of time-change—is technology saturated by one (or many or all) of the ways mentioned then there is no escape from technology. This does justice to paradise lost, utopia, nirvana, paradise, the Elysian fields. Perhaps the Tao remains, but strictly qualified as a media in search of its own message. Will the wo-man quest, after a cleansing of its idolatry of the feasible, begin to orient itself toward the conscious, willful action animating the inanimate in ever-increasing proportions? This animation will entail the creation of more and more minds coming forth from flesh or silicon or other alchemic brews—and thus entailing more and more consciousness. The futuristic theatre of such a surge is cosmos so crudely prophesized by science fiction. It will be a theatre of the absurd because connatural to the process will be the unknowability of its goals. It will be pure unadulterated eschatological folly with a secretive intent, with an end that I call the cosmic seed of evolution. It will be the triumph of theo-technology, the triumph of everything alive and kicking, made into the zero mass, zero space, zero time of resurrectional logos. It will be the final achievement, an all comprehending, equitable, beautiful acknowledgement of reality by its own self, the Omega Seed of becoming.

What then is the anticipatory scenario of hope in the short term?

1. A dropping of feasibility in favor of desirability.
2. A leaner and more equitable predisposition of things.
3. A sharper perception of the nature of reality, with the practical ultimately subordinated to the real.
4. A stripping back of intolerance and greed in order to see their clear and coarse pseudo-technological nature.
5. A parallel liberation of the technocratic mind so as to allow its sharpness to operate at the human logic level instead of the profit logic level. I make a distinction between profit and profitability: To make a dollar when bartering goods is to make a profit. The growth attained from reading a good book is profitability. We need to "integrate" profit and profitability.
6. A perception of the ease with which self-deception operates at the heart of both Capitalism and Marxism; and a recognition of the materialism that is eroding them both at the core.
7. A disconnection of technology from materialism to allow technology through desirability work out the metamorphosis of "matter into spirit".
8. A realization that in this transformative power technology will not only be explained and justified but experienced as the hope of life. For technology is the ship that is capable of taking life to its ultimate end—to spirit, by means of a process (evolution) that takes with it all of reality.

A recognition that the practical mind was finished long before it came upon this book. It was finished when reality was confounded by practicality for the first time. *The aim of life is the practicing of reality. The main activity of a practicing reality is the creation of reality. The creation of reality is fundamentally foreign to what passes for the practical in our contemporary world.*

Conservation/conservatism is one of the aims of practicality. It is to do tomorrow with some minor present readjustment of tools and equipment, exactly what was done yesterday. But reality is a process of transcending, it is a self-creative proposition endlessly on the move. Technology must develop so as to sustain this proposition. Present technology courts entropy. It is ballast instead of fire. We have a desperate need for a new realism so that we can ask from technology the things that will foster reality. Realism implies a recognition of ambiguity. It implies a recognition that the moment things become very clear is the moment the alarm bell must be struck. A very clear reality is a misnomer; an image might be clear, but what it stands for is a reality that is almost unknowable. Therefore, our quest and action for the desirable is fraught with a mystery only slightly dissipated by our fragmentary knowledge of the past and by our "anticipatory realism" of the future. With those two flimsy crutches we must do our best to develop a technology of transcendence, that will be a direct descendant of the technology of the flesh—via brain-mind.

The Western mode of civilization is right when it endorses metamorphosis. It is wrong when it acts as if metamorphosis was a hedonism of the mind. True hedonistic game-playing must be seen against the backdrop of fear of and microconcern for the relentless ramming of time at the gates of unknowability.

The "Eastern" mode of civilization is right when it scorns the conceptual fuzziness of Western frenzy for manipulation. It is wrong when it dismisses at least ontologically, the crucial and essential theological nature of manipulation-technology.

Begin to live (East and West, North and South) as if the most modest act of the mind or of the marketplace—regardless of its affordability, uniqueness, routine, or cost—is both an affirmation of concern and a declaration of indifference to the human condition. Anyone who buys in, or who is impervious to buying into the marketplace changes the balance of matter/spirit. Such balance is measured with equity (freedom from famine, from harshness, from destitution).

The mindless technology of "profit" is a symptom of a deeper problem. It is not a casual one. For to be a technocrat for profit, is to be one's own nemesis. It guarantees that one's own "personal" nemesis will pull down all those others those gullible, skeptical, envious, enraged, friends and foes alike. It guarantees that to the tons of inanities and faults there will be a corresponding despirited spirit humiliated by deception and greed but most of all lacking in a sense of worth and uniqueness, and in the fundamental aims and privileges of being-becoming.

On a personal level for the last ten years I have worked on a project that attempts to be coherent with the above norms. The project is Arcosanti which is as I see it, an urban laboratory in the making. It is meant to be a prototype for a learning/doing community of about 5000 persons. The pre-eminence of technology, in its larger meaning that I have tried to develop here is basic to the premises of a project that sees the urban effect as the drive within all living processes, and within any living process.

The central concept of Arcosanti is that the inception of consciousness within the cosmos may have triggered the beginning, albeit weak and insecure, of a willful evolutionary progression; and that intrinsic to such a progression is the process of the complexification of reality. The enormous complexity of the cosmos eventually pales in comparison to the immense complexity of the mental processes and the physiological mechanisms that support them. This process of complexification is labeled the *urban effect* because the city is a system, more than any other, that sees the transition of the already unbelievable complexity of the wo-man phenomenon into even more improbable occasions. These occasions are part of a political, economic, social, cultural, scientific, and technological totality, that is more than the sum of all its parts and is the epitome of complexity.

The urban effect is, then, proposed as the recurring and imperative effect running through evolution from the very early unicellular organism onward, and it is characterized with ever growing force by the implosion of a relatively indifferent milieu, the nonliving universe, into discrete, discriminately complex, and necessarily miniaturized systems. Therefore, the city that promises an access into the future must be a habitat that abides by the imperative of complexity-miniaturization-duration.

Because the urban effect is in all living things, it transcends anyone. And since we are living things, we find transcendence in the urban effect: that fundamental phenomenon in which two or more particles of physical matter begin to interact in ways other than statistical and fatal (the laws of physics).

The urban effect begins in life with not much more than a chemical sympathy between elements and achieves great power in the exploits of love. If the urban effect is a generic concept, it is so because it is a generative and pervasive stress that devours its children in order to reach for more self-knowledge and that integrity implied in its bonding power.

In primitive forms of life, the extension (size) of the associated systems (coral reefs, etc.) is determined mainly by environmental factors (moderate climate, etc.) The components (the coral polyps) are unconcerned about the extension of the association, because the association is deprived of deep psychoemotional ties, and it is also logistically loose. If the food is present, the climate is favorable, enemies controlled, and the building blocks within reach, that is all that is needed. The more we move toward complexity, the more the extension of the associative systems (bees, ants, humans) is determined by inner factors, because those inner factors assume greater and greater significance and power.

The coherence of the process through which the cell becomes the city is based on the exponential accretion of complexity by which information becomes knowledge and knowledge becomes the new mover for creation. This coherence cannot weaken until the ultimate condition is achieved.

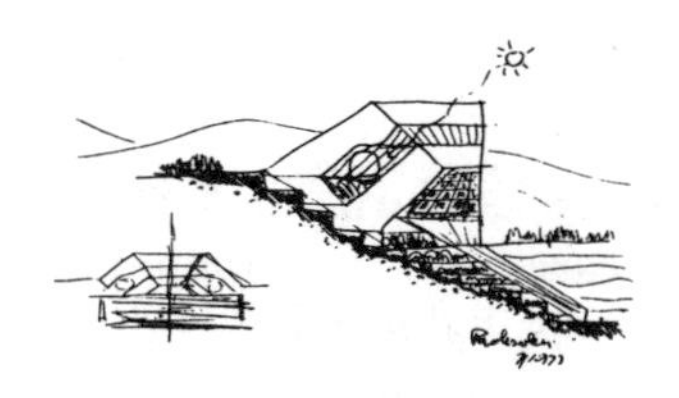

Chapter Two

A Methodology

The most commonly held mistaken opinion about my work is the belief that some years of "introspection" have produced a take-it-or-leave-it package solution to the urban problem. Rather I am proposing a methodology while at the same time trying to illustrate it. This methodology is original and radical in a literal sense. It is original in that it is "of and with the origin", and radical in that it is "of and with the roots". But it is the least original and radical when these terms are taken at their "market" value; because it is the oldest, the most used, and the most rooted of all methodologies applied to life. To grasp and to accept this methodology, one needs to be free of the Adam and Eve syndrome. That is to say, one must be free from the historian's mental block that makes one stop at the gate of pre-history on the tenuous contention that what preceded us was and is deprived of pedagogical, experiential, and methodological importance for us and is thus unable to teach us anything about our possible future, our mind, our society, or our culture. This is pretty unoriginal and unradical stance, untenable vis-à-vis contemporary science and unable to fit itself into the evolutionary momentum. This (static) historicism is hard put to explain how matter becomes thought and spirit, or how a fiery ball of mass energy, lost in a corner of a run-of-the-mill galaxy, has transformed and transfigured itself into the earth of today.

If metaphorically this incredible development is seen as a cooperative effort of myriads upon myriads of workers constructing their (collective) home by putting rectangular bricks above rectangular bricks (the methodology), and if among the myriads there has appeared now and then queer characters with extravagant round bricks that have been unsuccessfully and disruptively thrown together with their round bricks into heaps of rubble (into piles of mutants and their failures), then I find my niche among those myriads of advocates of the rectangular brick, rather than among the strange ones, those go-for-broke, opulent technocrats of all ages. Those among the former myriads are the rank and file. They began with the first, original and radical "simple" cells and proceeded up through successful mutations of all the species; they have come from the vegetable and animal, through the Garden of Eden into the demonic universe of the mind. They have travelled over those pain-filled bridge that are making matter into spirit.

It is when historians want to be pragmatic that they become pathetic. Their round bricks keep making a fool of them by allowing them some victories in a losing war of incongruence. They become, with the great technocrats, the staunch advocates of some better quality of wrongness, that true face of the polluted slough and its entropic backsliding away from the spirit.

As with any honest methodology, the original-radical methodology proposes a necessity, not a sufficiency. It does not suffice because efficiency is not sufficiently endowed to be an end in itself. It is plainly a good instrument for the constructing a bridge that pours more matter into spirit.

To our next question, "Do we need a methodology?" there is a pretty stark answer. We already *are* a methodology successfully developing itself into an aim. We are, when we are, free agents—inasmuch as, and only inasmuch as, we have methodologically invented ourselves from the first ambiguous living cells on, into what we are—quasi god-like automation, "freed" within the most subtle portion of ourselves, the mind, so as to prognosticate, and to plan for a less coerced future.

This methodology will deliver life and beyond as it has for eon after eon. And within its guidelines we will rediscover ourselves as a truly compassionate animal. This methodology is the process by which the early, vague interiorizations of physical bundles, (the first living creatures) grew in number, complexity, compactness and power throughout the evolutionary ladder until as persons, civilizations and cultures they came to be and became "routine" events.

This interiorization will not stop, if life does not stop. It will keep growing, if life keeps evolving. Arcology (the construction of "the ecological city") at its minimum best is but a mechanism necessary but not sufficient for such continuous interiorization of the world of mass-energy. It is part of a nuts and bolts pragmatism of matter becoming spirit, of hunger being overcome, of deprivation being cared for, of life becoming personhood. It is of the species, and all of the species staying on course; the course that has a future because it constructs the future. It does this congruously and willfully making horse-sense (the rectangular brick) of what it does instrumentally, freeing thus the spirit for ends unforeseeable because they are as yet uncreated.

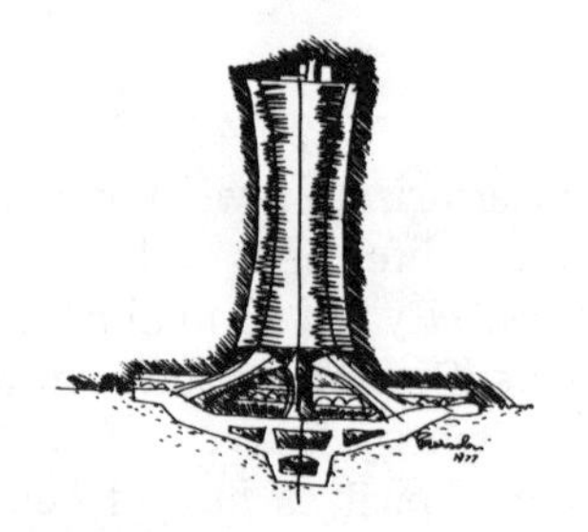

Chapter Three

Ecology as Theology

If a "model of reality" is a device by which knowledge "enacts itself", then the validity of the model, its truth, is proportional to the effectiveness it demonstrates via the pragmatism of enactment. If Euclidian geometry had not been useful to the technological enactment of the last two millennia of man's inventions, then it would have been true to lesser, limited degree or it would have been absurd. This is an undigestible notion, for so-called "pure scientists" and probably for two reasons. . . .The first has to do with the possible abstractness of their mental process which renders them incapable of making the connection between the statistical and the willful, between matter and spirit, since according to them the willful treads on "non-scientific grounds." The second might have to do with the "anticipatory gap" separating the model from the context of the present, a context unable as yet to see the pragmatism, i.e., the Truth of the model and consequently a context that sees the model as useless if not meaningless.

In the first instance, scientists are a pseudo-scientists inasmuch as their model of reality is intrinsically deprived of substance. The substance that makes life as real as, or more-so than, stone and fire. Their model is therefore a game of words, symbols, numbers, categories. . . .

In the second instance lies a case in point about the anguish inherent in the living phenomenon inasmuch as the most alive part of such a phenomenon is in a sense a pre-future, a cantilevering on the future, and as such liable to remain utopian, to be lost in the folds of a history engaged in other models. Therefore not only might the model be lost, but also its potential off-spring may never see the light of day.

If the "idea model" of the hand shovel had been lost in such a way, the steam shovel model probably would not have appeared. But even if it had appeared, the work accomplished by the enactment of the hand shovel model would have been lost. It would have remained utopian. Therefore, each model has its own historical "moment of truth" and unless the model comes to fruition in such time slots, the truth of the model is lost since its pragmatism (effectiveness) had no chance to develop.

The model developed in the thesis of the Omega God identifies the evolutionary (ecological) transformation of the reality close to us with a theological development. That is to say that a "God" is being slowly and painfully created; a potentially monotheistic divinity out of the utterly polytheistic reservoir of mass-energy and conscience. Of the many components in this process, the human kind is a pivotal element. For the sake and for the development of this pivotal element, there are post-biological inventions, the post-biological technologies, which are necessary, indispensable instruments to such further evolutionary inroads.

The Arcology, or "ecological city", is one such instrument. It is a complex instrument, but it is a medium which is in itself also a message, inasmuch as it is tangibly and pragmatically part of the transpersonal nature of life, (that part which is of a theological nature). In fact, ultimately Arcology must be stone made into spirit or it will be a simple mechanism of economic, political and logistical expediency. It has to be the spirit moving the mountains not so much because of the physical manipulation of the mountain (mining, etc.) but because of an inner flame elicited within the stone itself by design and grace. . .the process of esthetogenesis.

Then between theology and arcology there would be a "constitutional" tie; for this to be so it suffices that first, "*Theology is a genesis*" and secondly, that *arcology stands for a truly human landscape* that fosters such genesis. In other words in this "model" the psychosomatic man, the whole man, can be called the theo-arcological man.

First, *Theology is genesis,* in so much as whatever in the universe points at an increment of consciousness, points also at the further development of a sacramental, divine stand within the universe. (In Arcology an attempt is made to characterize the "theological" model as a progressive fulfillment of a "skeletally invented god", that is to say, the filling of a form-idea with a content-evolution, a true genesis of god.)

Secondly, *Arcology is a human landscape*, in so much as this immensely difficult and *hubris*-endangered genesis of the immanent god of the present (all presents) is to be a powerhouse where, by the virtue of such genesis, there is that which is inextricably co-active; thus it includes both the tangible (mass-energy, etc.) and the intangible (the spirit). And, that which is co-active within the complex containers of the physiological is also co-active within the containers of the social and cultural that go toward the making of human individuals as much as for the making of human societies.

For such a "demonic" creature, this totality of life and consciousness emerging from the mass-energy framework, creates a house in order. But this is a house that caters to real human needs, a piece of ecological architecture, whatever this implies at the various stages (presents) of its evolution. To invent the idea-city is to invent the house. The word "arcology" (architecture and ecology) is the term I use to indicate *one of its objectivations*: the (imploded) organization of the physical into a sophisticated instrument capable of *serving* and *ex-pressing* what might be called the transpersonal consciousness of the society, a cultural-theological-creating genesis. *Serving* inasmuch as it gives to such society the physical structures for its actions. . .*expressing*, inasmuch as such physical structures are for better or for worse representative or, in more germinal situations, anticipatory of its ethos.

In other words, arcology is in the human landscape, if in the genesis of the humankind both matter and spirit are present as a continuum. Then today, as any moment takes on the durational process (any day of history or pre-history); and it documents a more or less specific and successful co-ordination of matter (soma) in such a way as to cause matter to transcend itself. Most inner light is the spirit (psyche) of co-ordinated matter. Or better still, most inner light is the spirit of the co-ordination of the physical in specific and successful systems.

But assume, as it is assumed here, that humankind is more than the sum of its single persons. Then this "more" needs a locus and a consciousness (a soma and a psyche). One locus and one consciousness would imply a "fulfilled model" of the human experiment. Thus it is more realistic to expect that fragments of locus-consciousness will be the pluralistic foci for this reality; the reality where the person is more than itself.

Historically the typical and most dynamic foci was the city. I choose to call the city of the future, "arcology" simply to remind one that in human terms a good city is an "architecture of ecology". This simple fact bears radical fruits; bitter fruits for individuals who believe the world to be their private hunting ground; and exhilarating fruits for individuals who believe there is a theological component to their significance.

Indeed after an encounter with the theological dilemmas one can firmly state that if a theological component is the backbone of the evolutionary phenomenon, then arcology (the city) is necessity on its way to virtue. Some mystics may argue that if one eliminates the possibility of attaining the spiritual condition by non-material means one bestows too much on matter. In debating this one should consider that if matter is rejected as a basic, necessary and indispensable component of the spirit, then one is faced with the unavoidable senselessness of an evolutionary tide tragically involved in transforming "material" into "immaterial." Without such a theological backbone, arcology is rather just a necessity dictated by the laws of mass-energy. This is so for those who see no reality in spirit-imbued matter. William I. Thompson in his book *Passages About Earth* says,

> Evolution moves against the direction of entropy; as matter moves toward more probable states of maximum molecular disorder, life moves toward increasingly more improbable states of maximum molecular organization. More and more is packed into less and less, until the miniaturization process reaches its greatest level of what Teilhard calls 'complexification' in the compactness of the human brain. The simplicity of its size and shape belies the dazzling complexity of its interior. For this process of complexification linked with miniaturization is the lesson the city planner should take away from the study of nature. In evolution, simplicity is always linked to complexity: while huge dinosaurs lumber into extinction, tiny mammals chatter in the trees. I would say it is much the same with our cities now. The huge megalo-

> politan beasts are sprawling all over the earth; in terms of thermodynamics, they are spreading their energy equitably through space and approaching the heat-death of entropy. They destroy the earth, turn farmland into parking lots, and waste enormous amounts of time and energy transporting people, goods, and services over their expanses. They so fill their ecological niche that they destroy it, and thus become caught in their own evolutionary dead end.
>
> Soleri's answer is urban implosion rather than explosion. The city should contract and intensify, but in order to hold its information in negentropic form, it should imitate evolution and complexify itself through intense miniaturization. A city of 600,000 should become a single, recycling, organic arcology. The people would not live crowded in ghettos but on the outer skin of a towering arcology that faced toward a nature that was once again natural. Thus the surface of an arcology would be a 'membrane and not a wall.' Inside the arcology, along its central spinal axis, would be not the natural but the civic space. Here society would turn inward for the concerns of man and culture.

If such a theological backbone is only imaginary, everything has the same worth, is both necessity and virtue; becoming the empty frames of a mindless reality.

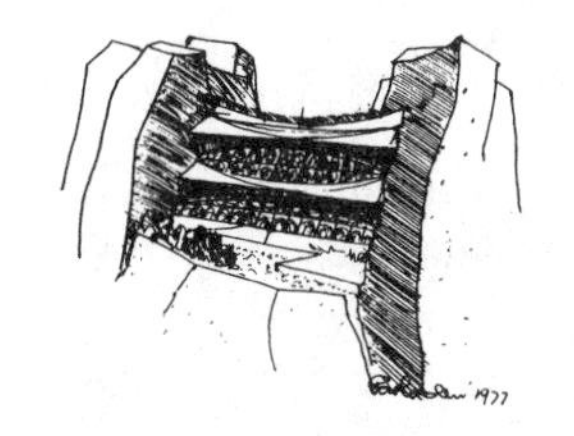

Chapter Four

Wo-Man—Space—Justice

To explain the reasons for a space colonization program, I would start with a parallel between a revelational model and a creational model of the universe. Within such an ontological or theological parallel, I would set immanence, (the present) and from there proceed to justify space colonization. The justification would be eschatological.

Revelational Model

In considering the revelational model, one can propose a variety of metaphors whose common background is a clear or hidden symbol of animism. For instance,

The cosmic landscape metaphor

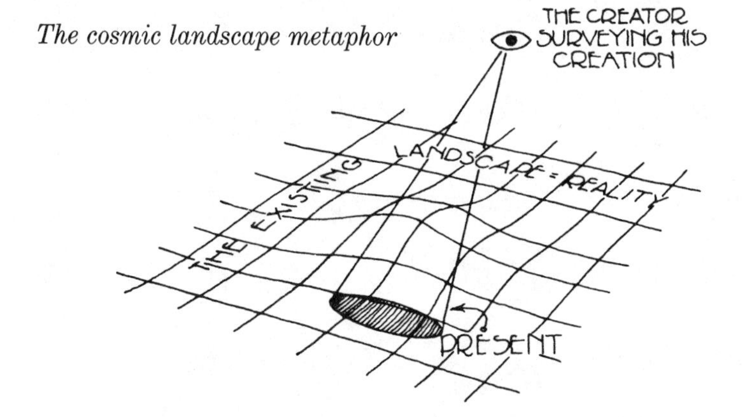

The present is the scanning by the creator of the existing reality that (s)he originated. Time is the beacon of light focusing on successive and forever-existing landscapes. It is a fatal universe, established once and for all, generated by a relatively leisurely mind and in part justified by appearances.

The cosmic jigsaw puzzle metaphor

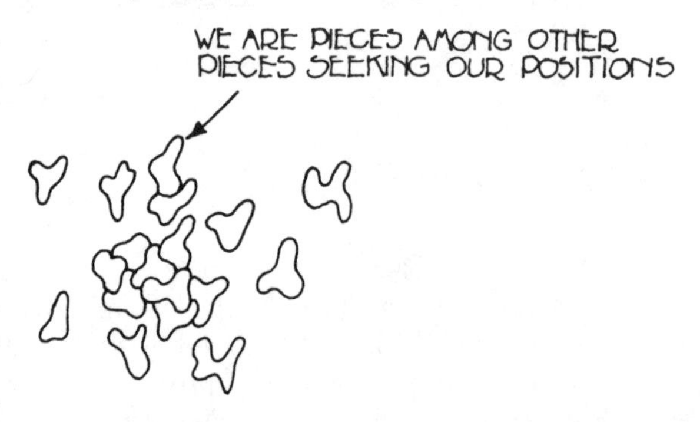

Reality exists and is operational but given the limitations of mind it is perceived like a scrambled jigsaw puzzle. Unscrambling the puzzle will reveal the full divine design which pre-existed the mind and will allow the protagonist to re-enter it. The creator can be the puzzle or its originator, a fated universe generated by the need to fill the eye's mind with something to counter the physical emptiness of the desert (which tends to fill up with mirages and hallucinations).

The void tortilla metaphor

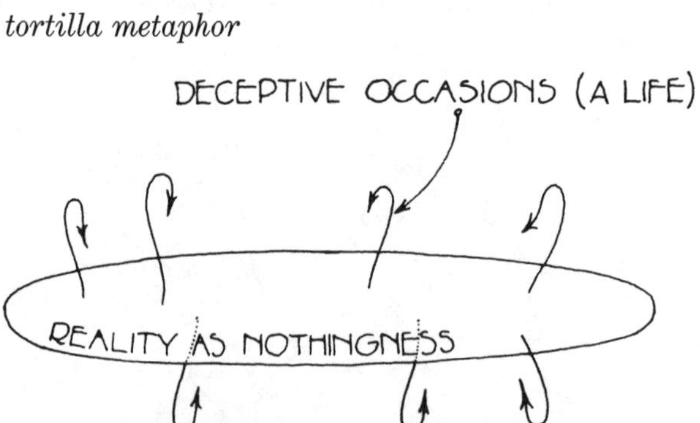

DECEPTIVE OCCASIONS (A LIFE)

REALITY AS NOTHINGNESS

The only reality is nothingness, the void of the Tao or a nirvana. To achieve oblivion is to be blessed. Reality, as sensed by wo-man, is composed of deceiving occasions. A fatal reality, unalterable and limitlessly pessimistic.

This view quite possibly originated in a society in which suffering was endemic, unbearable or inescapable.

The graffiti metaphor

Reality is love and love impregnates the universe which grows in it. Reality is benign energy, a given and giving indestructible radiance. This perceptive is fatal, and dogmatically benevolent. It is generated by immature gullible and arrogant minds.

The hot-air balloon metaphor

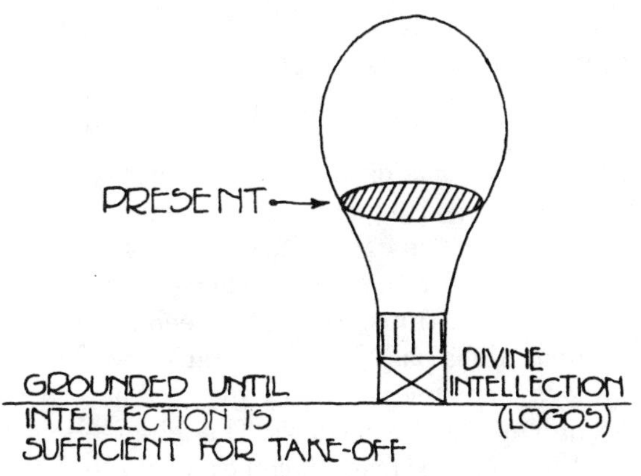

A divine intelligence is at the origin. For this to "fly", a process and a technology are instituted. The mind-full canister takes off when the process has enough momentum (hot air). But in the first place a non-flying intelligence is also non-intelligent. And coarseness needs no refinement to explain its existence. Such beliefs are fatally flawed, generated in minds that are science-oriented but mystically-driven.

One can scramble these models and produce variations, other metaphors. Then there is the *model of despair.* Any metaphor that points at senselessness would fit since despair is generated by the perception of an ontological meaninglessness which automatically makes reality the cruelest of hoaxes.

The Creational Model

The super rubber ball metaphor

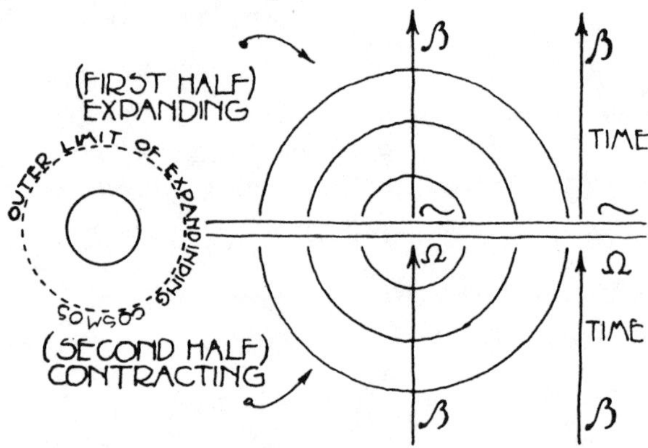

Reality originates as mass-energy, time-space. All along, through its expansion (Big Bang) and contraction reality creates itself by consuming its own agents, mass-energy and time-space, into spirit. In advancing from Alpha to Omega, the seed of all the "processed" reality that transpired in between makes the difference. The original Alpha has, therefore, transcended into its own entelechy, Omega, and Omega is truly the seed of a cosmos that has gone through *total* esthetogenesis. The self-creation of divinity is not predetermined and may not come about. If it does, Omega Seed is resurrection.

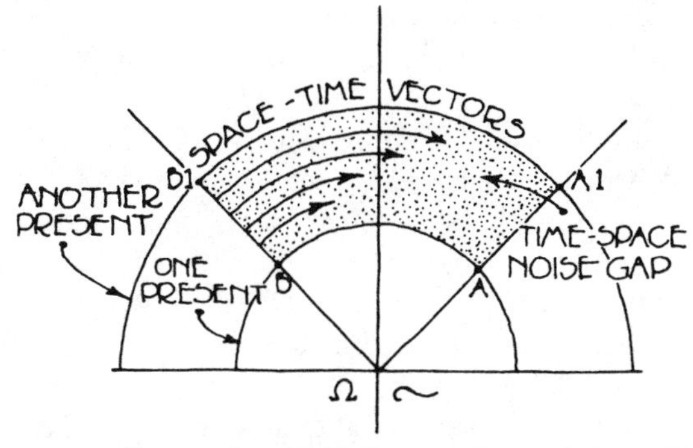

The expanding and contracting surface of the rubber ball defines the present (understood as the contemporaneity of events and occasions). The sphere is not tatooed (see the cosmic landscape metaphor) from the very beginning with a divine design which will grow or contract with the sphere according to an immutable pattern. The sphere is originally blank and is painstakingly self-tatooed by the creational process. While an observer is co-present with all the events displayed on the surface of the sphere (as present), the same observer sees only events of the past just arriving now by way of light or other types of messages.

For instance, as the sphere swells, observer A, contemporaneous with observer B, will perceive B after time interval X, the time necessary for B to "send a light message" to A who by then has become A′. There is no way by which A and B can really coexist and know each other while coexisting. We never know what is contemporary. We have access only to "ancestry." Since a space-gap is separating A from B, a certain time must pass in order for A to perceive B. A will perceive B from position A′ at distance B′A′. But time is change. Therefore, A′ cannot be equal to A and B′ cannot be equal to B. (All this is for A to take a glimpse at B.)

The moment (occasion) A′ is on the traveled path of occasion A (a person at two different moments of life) and radiation B′A′ is one of the infinite radiations spherically originating from B. In the time interval B′A′, occasion A has evolved (aged) into occasion A′. Therefore, A will never set eyes on B, his contemporary. It is the progeny of A, one of which is A′, who will see or encounter B. By then B is B′. Therefore, perception, understanding, or knowledge cannot ever refer to contemporaneous events but only to past events. The present is inscrutable to itself. A will never know B and vice versa unless. . . .This is the one single physical fact which makes any claim about absolute truth, perfection, harmony, justice, or love totally gratuitous.

I am now (and there is no other me). I can only hope to know the past, that "which is no more." My contemporaries are inscrutable because they are unreachable. Time must pass for B to reach A (and vice versa). Even if the message of B does not suffer the noise of time-space, A and B will become A′ and B′. The coarseness of mass-energy and space-time is the obstacle to an occasion knowing its contemporary.

Only a radical transformation of mass-energy and space-time can alter such an endemic and inevitable situation. For instance telecommunication, is impotent on this score since the speed of light is the best technology applicable to its performance.

The web of predictable behaviour, the web of culture, and the determinism of matter sustain a situation otherwise altogether impossible. But all three of these belong to the conservative (past) side of the spectrum and must be constantly overhauled and overwhelmed by the new, the creational.

There is no need to look at cosmic phenomena to verify the above. It is quite apparent between two people talking: by the time the response is formulated and spoken by one, the other has his mind elsewhere. In addressing interlocutor A, you actually reach A′ who in turn sets up a response which will reach not you, B, but (s)he who will be B′.

If I do not know my neighbor, I cannot be just with him, "honest" with him, in harmony with him. Nor can I combine with him to achieve perfection, even assuming the optimistic and purely theoretical position that all possible knowledge would be applied. But for the separation from my contemporary to become zero (assuming for the sake of simplicity that communication is equivalent to understanding and consequently disposing of the obstacle toward perfection, justice, grace, beauty and so on) all distance must also become zero. Justice, grace, beauty, etc., can exist only in a dimensionless universe. This is the imperative of Omega Seed, the center of the rubber ball metaphor.

If any interval (for instance, a time-space gap like B′A′) becomes zero, all intervals dissipate into zero because all intervals are fractions or multiples of any interval and fractions or multiples of zero equal zero. "Infinite", more than immense, miniaturization is mandatory in the structure of the Omega Seed. Time never rests or waits and covering space (the transmission of information) takes time. Since time is change, change is intrinsic to any interlocution between occasions. The targeting is, therefore, necessarily imperfect since in the best of circumstances we must decide (but don't) between responding to that which is no more (A) and that which is not yet (A′). The utter ambiguity of becoming could not be more evident. The paradox and irony is that without the obstacles of mass-energy and time-space, that is with the disposal of ambiguity, one also disposes of all occasions: no A, no B, no A′, no object, no subject, nothing to kick around anymore. It is like a hurdle course: the hurdles interfere with the run but the run cannot take place if the hurdles are taken away.

This destroys capital "J" justice, as well as perfection, beauty, grace, etc. But there is far more to it than the impossibility of present justice. Total justice is retroactive justice. And this probably only gives us two alternatives:

The past is irretrievable; justice will never be.
Justice must be; the past must be retrieved.

To look for other alternatives is a futile exercise if time is to have its due (see the rubber ball metaphor). If time is pure expediency played with by a timeless being then it is up to such a timeless being to render justice. Such justice must carry some meaning, otherwise it is a mockery of truth since its truth is a mockery of justness:

1. *The lack of knowledge of the past, the need for its understanding and retrieval.*
2. *The inscrutability of the present and its hammering out of reality.*
3. *The nonexistence of the future and its anticipation by intellection and will.*

The conscience of life has to cope and work with these three conditions. To the hermeticism of the genetic code, a one-way command, is added the hermeticism of a present (all present) rendered incapable by the nature of things (mass-energy, time-space) of communicating with itself. It might well be that just because the present is inscrutable to itself, the past, incorporated within the genetic code, must be dogmatic, indisputable. If it were not so, the nongenetic guidelines might turn out to be insufficiently strong (may be extravagant) to carry the day. A domineering genetic code is like a large wheel with its circumference bridging the potholes that plague the pavement of becoming. The larger the wheel, the more single-minded the genetic code, and the more possible it is to run over very rough ground without undue cost. But if the wheel gets to be too large, insensitivity sets in and with it the risk of fossilization: scorpions, termites, etc.; in other words, the potholes are challenges to be dealt with, not gotten away from.

If the preceding is reasonable, the norms can be deduced. Space colonization is the generation of centers of consciousness and action in places other than earth. It is an urbanization of the cosmos at hand. Urbanization is the most direct and powerful mode of dematerializing the mass-energy, time-space reality. *Every living phenomenon, excluding none, is an Urban Effect, miniaturization in the discrete context of the B′-A′ gap.* The gap B′-A′ is not simply a derangement of time-distance. Sealed in such derangement is the nature of misinformation, misunderstanding, and its consequences of estrangement, ignorance, dread. . . aggression, violence, sufferance, injustice, and so on. As indicated above, the dilemma, the existential anguish is that a measure of "derangement" is necessary to guarantee the presence of the media needed for transcendence (creation). The media, mass-energy, and time-space is thus both culprit and resource.

The Urban Effect is therefore an imperative, quite possibly the imperative of a process that wants to signify itself. *The process goes on to its 'logical' conclusion. Its signification becomes its justification, its advent of justice.* The advent of justice is a necessary corollary to the advent of grace, an esthetogenesis of the cosmos. Then the space program becomes necessary as an urbanization program: the cosmos injected with mind.

Technology is a two-edged sword. It is a carefully-controlled instrument that can cut into the indifference of matter and deliver wo-man to consciousness. To a refined neo-matter that advances the creational process. But technology when it is an uncontrolled frenzy of over-simplification cages wo-man in its deterministic vise;

then it crushes and kills wo-man's spirit. One can easily foresee monstrous space cities dedicated to the technocratic idols that promise more and deliver less of what we call grace.

Such a specter cannot halt our quest for grace. But this quest is not to be pursued in the Elysian fields of conservationism or simplification. It must be carried on within the disquieting center of urban turmoil, and in resonance with the protracted miracle of evolution.

This is the complexifying, miniaturizing and becoming of divinity. In an endlessly protracted but inescapable way, the quest for justice, truth, grace and love, and the expansion of human action throughout the cosmos are inseparable. The fulfillment of one is conditioned upon the achievement of the other. To say it once more, this inescapable dependence is in the nature of reality. Every step toward the spirit is a transfiguration of a portion of matter that is necessary for that step. Eating is indispensable to thinking. A wave field is indispensable to music-making. Materials are indispensable for building the temple. The cosmos is indispensable to the creation of Omega Seed. Since Omega Seed is absolute (or is not), the process must be "infinitely" involved and involving. How else can the absolute of justice, beauty, and grace be created?

Retroactivity, the knowing and loving reenactment of the past, can only come about through a resurrectional miracle. The miracle is exactly the same one that evolution is engaged in: the consumation of the mass-energy and space-time needed (when A+B=A′) for the creation, total and ab aeterno of divinity. By then the A's and the B's will be immensely living and loving. . . .The imperative of justice-love-grace and the imperative of space exploration are inseparable. The case made here is not for instant communication A−B=0 per se, but is about the intrinsic "defect" of a reality in which information is inescapably distorted. It is misinformation. It is furthermore assumed (inferred?) that a world working at the reduction of misinformation is an imploding world in agreement with the notion of an ultimate "divine center", all-informed and all-comprehending, knowing, loving, beautiful; an estheto-gene-sis.

Space is a necessary but clearly insufficient condition. *How* we go into space is the critical question. Here I mean not so much its mechanics but its intent. If we have ever needed true saints, this is where we need them. Space saints. This will disqualify most of us. The cultivation of saintliness must be a priority for the space venture. The saint-scientist must be one with the techno-artist; someone with a lean mind and a frugal body who exudes justness, economy, compassion, gracefulness, tolerance, understanding, trust, commitment, and self-responsibility. . . and who knows how to smile.

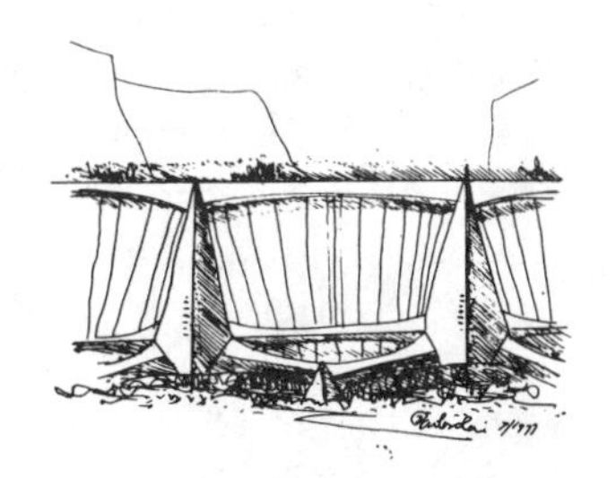

Chapter Five

Art and Environment

The art-environment debate is well-shredded and perhaps beyond redeeming. In this chapter I will attempt to forge a new linkage by proposing some unusual characteristics for a union of ecological, theological, and esthetic elements.

Resistence to this chapter's proposition will be minimally of two kinds:

1. The premises of the proposition are arbitrary with deductions poorly founded and shaky.
2. The proposition may have merit but good artists do not need its intellectual gymnastics.

But I will hold to the proposition because it offers a *modus vivendi* which does not ignore "facts" and partially at least explains why life is not a bed of roses (without making it desperate or senseless). The proposition is as follows:

First Premise: Since the inception of wo-man the environment has been mind-full. The human mind is anticipatory, seeking to define goals and ends. Ecology is thus a *mind-full-end-oriented* process; it is an eschatological process: theoecology. Goal-orientation is a mundane version of eschatological motivation.

Second Premise: Religion is anticipatory thinking which produces hypotheses and then constructs models to "verify" them. Religion is wo-man's most comprehensive simulation exercise.

Third Premise: Art, the esthetic, is the transcendence of immanence through creation. This transcendence may be the only option available to consciousness that has emerged from the otherwise unabatable anguish of the species. In small doses and sustained intensity, the esthetic achieves what religion anticipates through its paradoxical models.

> Therefore, religion, acts through ecology. It is the anticipatory continuum dotted by bright spots of achieved grace. These are esthetic (creational) acts. This perspective places art at the very center of the ecology-environment-theology interaction.

We are then dealing with a triad:

1. *The Environment*: an immanent manifestation of an eschatological drive (consciousness).
2. *Religion*: a simulation model of anticipated perfection and grace, with a concluded reality of a fully-aware divinity.
3. *The Esthetic*: an emergence of spirit from matter in specific particles of grace (a novel, a dance, etc.).

> The process of reality creating itself might be called esthetogenesis, the transcendence of the mass-energy-time-space universe into grace, the beauty of a gene (genesis) imprinted with the whole cosmic process of self-creation. The indispensable "function" of the esthetic is predicated on the want of grace in a non-animistic universe.

In an animistic universe, creation is simply the revelation of an already-existing grace. This revelation is not pivotal since that which is revealed exists *a priori* and its revelation is non-essential. It might be ego-gratifying that the prime mover reveals itself to wo-man, but the act is somehow hollow. The prime mover, perfection, has no way (even if willing) of sensing the imperfect inasmuch as the existence of imperfection would denounce the existence of perfection and simply dispose of any perfect, immutable, loving *causa prima*. This means that inasmuch as imperfection exists, perfection does not.

The fantastic theologies and their incredible models which enrich history are what remains of the enormous efforts made by wo-man to rationalize or mysticize the dualities that the mind senses as being crucially lodged at the heart of its own meaning: the dualities of good-evil, everlasting-temporal, matter-spirit, and so on. Science has deflated most of these models while engaging in the pursuit of its own elusive nature. Now, as the scientific myth is paling and often self-punishing, the mythology of religion dissolves into esthetic fragments and morality plays. God is dead inasmuch as "he" never existed as a mature "phenomenon". And if God is not, then the duality, God-Wo-man, with all its mesmerizing contradictions, may be conceivable but is improbable.

If, on the other hand, the pristine universe is non-animated (science says it cannot tell. . .well, perhaps religion can) then any speck of potential or incipient animation counts. There is the animation of ecology and the animation of the esthetic: the evolutionary continuum and the discrete, creational, esthetic act. Religion is the glue, or better, the plasma uniting the two. That bonding agent is anticipation which is given design each time a religion is founded. The design, loosely or closely, is anthropomorphic, the everlasting in the image of the immanent. This is perhaps the reason why all the anticipatory imaginings of wo-man are in the end ridiculous or macabre—unless buried in a work of art (Fra Angelico, for instance). Nor are the symbols of fire or light, the geometry of mandalas of words or numbers really telling of that which it is impossible to imagine.

Instead of trying to portray or describe it, it might be more useful and less arbitrary to speculate about what it might stand for, or better, what it should stand for. We could say grace, perfection, beauty, justice, love, knowledge, imperviousness to degradation. . . inclusiveness. . . In a way each of these attributes *is* all the others combined because a default of one automatically cancels the rest. All circumscribe one phenomenon: the cosmos making itself into logos. If we lump these attributes together and call the bundle Truth, then the pursuit is in double jeopardy: it is difficult to discover the past, i.e., fragments of truth; and, it is impossible, nonsensical, to know future truth, to know the non-existent. *Truth does not exist.* This is not nihilism; it is that horse-sense of reality in the process of creating itself.

One thing we can do is to reduce any contradictory factor to a minimum, so as to establish the plausability of potential truth. This does not refer to feasibility as such, but to the necessity of disposing of self-contradiction. And it goes for all the other components of an ideal reality. We are dealing here with absolutes, not 10 percent, or 75 percent, or 99.9 percent, but 100 percent.

If we agree that imperfect perfection is no perfection at all, then imperfect love is not love, imperfect beauty is not beauty, imperfect justice is not justice, imperfect truth is not truth, etc. Incompleteness is by necessity perfectable; therefore, incomplete beauty is perfectably beauty and so on. Then our speaking is always hyperbole. We do not mean that something is beautiful. *We mean that something, weakly or powerfully, points at something ineffable and remotely possible which we call beauty.*

This is important. It tells us about something we can do by telling us what we cannot do. We cannot, for instance, lull ourselves in the expectation of the millenium. It tells us comprehensively what immanence keeps hammering in our consciousness. We are the barely-conceived embryo of a possible, if remote, grace. To allocate to ourselves more than the very humble tasks of brick-makers and brick-layers is pretentious, even if clad in modesty and love. We are elected not to be the elected by the very nature of that of which we are a part.

There is another incredible message in the notion of an attainable, if remote, all-comprehending grace. It is the notion of resurrection—because there is no way for the all-comprehending and graceful to exclude anything from itself. Consider for a moment the plausability of a comprehensiveness excluding something, anything: can grace *be* if even one disgraceful occasion exists and stands separated? All "disgraceful" occasions must end by partaking of being, of grace. Only a simultaneity of occasions gives all occasions access to the true measurement of *the* event: the flooding of all occasions by a radiance of grace, and an acknowledgement of being part of all.

In resurrection there are elements of redemption, penance, salvation, and atonement, but most of all there is a transcendence through a suffering survey of all occasions involved. Through this knowledge, this omniscience, love becomes concreteness, the conclusion of a process that turns out to be the esthetogenesis of the cosmos.

To Expose Religions, Art, and Media Peoples to a New Kind of Environment with Its Own Specific Challenge and Potentials

Scientific knowledge and technology have opened a whole new way of seeing wo-man in relation to the cosmos. It is imperative that religions, art, and media people attune themselves to some of the implications of today's scientific models and connect by way of them to the surrounding world. Arcosanti attempts to have its "ear to the ground" physically, theologically, and esthetically. It should therefore be able to generate in religious people, art people, and media people a passion for grace and expression.

To Integrate Performance, Habitat, Environs and People within a Media-Drunk Culture

In order to create a synthesis of religion and art, nothing will suffice short of a new and true economy where the ordinary routine of life is surrounded by the extra-ordinary. The aim is town as learner, composer and performer; with the inner resources of the community acting as both an excellent source of self-generated events, and as an optimum resonator for works produced by others. The community is made up of people, of institutions, and of a double character environment: the manmade and the natural. The town and surroundings can eventually become a home, a manufacturing plant, a cathedral, a theater, and a castle all in one, filled with stresses, tensions, and yet. . . and expressing, even creating.

Media is the electronic "intruder", it is the time-space sorcerer offering new dimensions and new avenues to the creative process (instant replay, memorizing, editing, montage, recollection, abstracting, bridging, interconnecting, documenting, intensifying, pacing down, speeding up, satirizing, mocking, symbolizing, shocking. . .).

To Reconcile Esthetics and Theology

It is traditional for madness to be placed on the edge of both prophecy-mysticism, and artistic prowess. Through my eschatological hypothesis, I see the prophet and mystic as powerfully anticipating a condition of grace, and the artist as producing the fragments of such grace. It is traditional for the artist to construct sacred images and sacred spaces on a blueprint of gods invented by religious leaders. This symbiosis has produced marvels. Given the new insights of science and the new tools of technology, a quasi-virgin (if only potential) cosmos is now open for the unimaginable and the graceful. For this to occur religious wo-man must again meet esthetic wo-man and find new grammars with which to write new poems to life about life.

To Snatch Leisure Man from Sloth and Mediocrity

This country is drifting year after year toward expanded leisure-time and sloth: shorter working hours, more wealth, more time and attention has turned to "pastimes." No spots on the American map more sharply symbolize such pastimes than Las Vegas and Disneyland. The former offers fun and extravagance peppered with thrills of legalized lawlessness. The latter offers an ever more saccharine treatment of the caring and the brutal, the simple and the complex. In both clever entrepreneurs have taken storm clouds of nostalgia and turned them into a rain shower of dollar bills.

Recreation in a Las Vegas–Disneyland mode is, for the most part, passive. Participation is a stack of chips, a book of tickets, an artificial environment that aggressively saturates the senses. Little personal involvement or commitment is required, merely presence.

America needs other *identifiable* options. It would be a privilege if Arcosanti could be such a center of such options not only for Americans but for others as well.

A commitment to critical mass and excellence are necessary ingredients not only of a new symbol, but for concrete alternative that will be a welcome contrast to Las Vegas and Disneyland.

What leisure wo-man becomes is a matter of critical long term importance. What looms ahead of us is a mix of transcendence and mediocrity. It will be tragic if mediocrity gains the upper hand and sloth becomes the character and end of the post-technological era. It is very important that the Las Vegas-Disneyland-Arcosanti triad reality be seen.

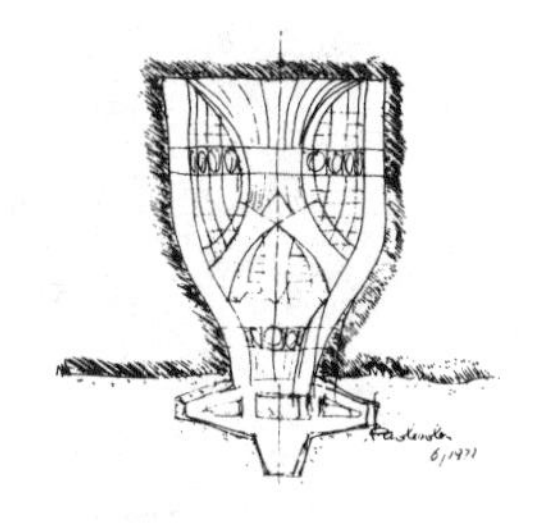

Chapter Six

Reconciling Esthetics and Theology

Brittle scholarship prevents me from knowing whether and why the esthetic sense proceeded or proceeded from the religious sense. But since I have strong opinions about the nature of the esthetic and the religious, I will share a few notions.

To begin with, some of the components of our early wo-man's world must have been awe, curiosity, dread, anticipation, and a will to live. Some ways that living might have been undertaken might then be schematically proposed:

1. Keep the components at a distance by sticking to those practical matters of living and developing technologies suitable for survival and contentment.

2. Lose oneself in a tidal wave of emotions and remain transfixed in adoration of a "tremendous mystery" (mysterium tremendum); seek only after redemption and salvation.

3. Keep cool and apply the analytical mind to deciphering bits of objectivity (science).

4. Touched intellectually and emotionally by the tremendous mystery, to go about creating bits of mystery of one's own, by making spirit out of matter.

Contemporary jargon could define...

> The first way as technological with the subtitles of economics, politics, and management.
>
> The second way as religious with subtitles of mysticism, animism, fatalism, etc.
>
> The third way as scientific with subtitles of mathematics, astronomy, physics, etc.
>
> The fourth way as artistic with no subtitles.

Remaining with our theme of esthetics and religion one could point to a difference between religion and esthetics. The religious person "contemplates" the mysteries, and the esthetic person makes mysteries.

The religious person is engaged in an uneven encounter between the discrete (the I), and the limitless (the Divine). Because of such a crushing imbalance, *a priori* surrender becomes imperative. "Oh, God, I trust myself unto thee."

Animistic reality gives witness to the relationship of the self with a divine presence. Revelation then is the true vehicle for salvation (or redemption, or everlasting bliss, or the love of the *causa prima)*. And reality, the truth, is revealed to consciousness through direct teaching, reflection, learning, and scientific discipline.

On the other hand, the esthetic wo-man is called to be a receptacle of what pours from within and from without the self. And through a mysterious chemistry to create that which is beyond description: bits from the art work of tiny bits of divinity.

In religious wo-man dread and awe are routine. But what is more consuming is anticipation. Being and becoming bridge the gap between the discrete and the limitless. But perhaps anticipation will be more all encompassing than that, if we contemplate the possibility that the concreteness of divinity itself might be no more than a wish so powerful as to interpret that which could some day be, the divine, as that which has always been. If so, religion in this sense is simulation and possibly a self-fulfilling prophecy.

In esthetic wo-man the unrelenting master is anguish. For an artist, deprived of faith in a promised land, to act means to produce that grace which is limited and crude; thus the esthetic act makes dread and awe even deeper. This I will never bridge the chasm between itself and the limitless, a limitless that may be an I expectant only of its own creation.

Saints are the harassed strategists of re-entry into grace who design fabulous blueprints of both vehicles (wo-man's strides) and terminal stations (God's icons).

Artists are the desperate tacticians sweating away in an endless toil of constructing (creating) on blueprints that are so ambiguous that they force them to improvise. The religiousity of the artist is more like a dike between the last work and the next true, believable bridge to salvation that comes along. It keeps the torrent of anguish from flooding over everything. For the artist, salvation is only a misnomer for creation.

Perhaps for the saint, this doubt and dread can be similarly contained if God's beauty and creation can be made tangible. By the artistic process, anticipation forces simulation in the mind's eye of the saint, and the artist becomes the instrument for it.

Therefore, while the artist turns to the saint in moments of high anguish and low inspiration, the saint, also when his or her own prophecies become too abstract, turns to the artist for immanent tangibility and credibility.

Since through different perceptions they both point to some transcendence, there might be a dovetailing of the two in an incompletedness that would produce an immanence of greater richness and intensity. Here the perception can originate from opposite notions of reality. On one side is the revelation of reality. On the other side is the creation of reality. The fires of controversy might take its partisans into the most unforeseeable places.

Religion and esthetics are two different kinds of threads crossing one another again and again, at times with lightning and thunder. With this love/hate duet, it is difficult to guess whether we are on a divergent or a convergent course. We do not know if a tandem journey is possible or desirable. Since there on a parallel path and to one's own horror, the road might go in two opposite directions: revelation for the saint, creation for the artist.

Even for the least focused perceiver the dialectic remains just that, a dialectic. Is truth parcelled out to us from a universe saturated in love? Or are we witnesses to the creator of truth being born in the womb of the universe?

For saints and artists to ignore each other is not only unwise, it is impossible. Would a competitive escalation of their common reverence for life produce a new, fruitful synthesis?

Religious fanatics want to save souls even at the cost of their lives. Fanatic artists accept ostracism and obscurity in order to remain faithful to their passions. There is an unmistakable difference in the way each relates to others. For the religious fanatic, universality means the conversion of the whole world to the truth. For the fanatic artist, universality means the intense concentration on a particle act or object so as to transfigure it into a universe.

Religious fanatics want to see manifest in others the same other-worldly passion manifest in them. Artists want to objectify the earthy passion that possesses them, no questions asked.

If I lean strongly toward an artistic model, then to reapproach theology by means of esthetics means to persuade saints that what they really talk about and seek is the progressive creation of a beautiful and ultimate condition of grace rather than an unfolding by way of mystical or scientific revelations of that which pre-exists in the lap of a divine *causa prima.* Perhaps because of the awesomeness of the two great contenders, I have little patience for lesser ones such as the artist-therapist, the artist-hedonist, the priest-psychiatrist, or the priest-exorcist. These are morality plays in the gray area with a priest who wants to create and an artist who wants to convert. They are important, but do not shake the world like the fiery prophet or the "artist maudit" do. So perhaps the real approach involves the acceptance of the interaction of both these two well-defined spheres, revelation and creation and for the time being letting the chips fall where they may.

Chapter Seven

A New Kind of Environment

If it was true that science and religion, while pointing at each other's paradoxes, are cornered into uncomfortable postures such as: religion is irrational because its laws are incapable of proof and measure; and science is the slave of determinism and is consciousness negating its own concreteness; then one could suggest an alternative of a time-loaded model. *The world is moving from the reality where science is most comfortable, a non-living universe, into the reality that religion has been anticipating for millenia, a divine universe.*

On the basis of such a model, science might wish to be less dogmatic about unchangeable yet quite unscrutable physical laws and admit that not all effects are written in their causes. Religion might wish to confess to a guess-work of a most anticipatory kind in its inventing and proposing of notions as if they were divine laws. They both work on the past. Science works by analysis, deduction, and inference of things. Religion, by what may be called a twist of fate, since the past that religion is revealing is quite possibly the future that religion is anticipating.

The time-loaded model is eschatological. Time has a beginning and therefore an end. Environment then is change, or time. It is not a spin-off from the grinding out of cosmic laws nor is it a better-than-nothing extravaganza of under-employed dieties. It is the very context of the process where matter transcends into spirit. It is within this frame that I see it as imperative for science and religion to carve out their niche within the ecological process. I call this carving out "esthetogenesis" since in it means and ends ultimately coincide; and with a result beyond description since it is creation, is beyond description. Protracted enough, this creation can provide science with an ultimate and all encompassing truth. A truth that finally can provide religion with that excruciatingly dreamed of and desired grace: the omniscient and all-loving Divinity.

For the time-loaded model, this frame provides a pure duration that is tantamount to resurrection. The resurrection within it can be called the seed of the universe, the Omega Seed. This is an ending that is symmetrical to the beginning, the alpha media, if it were not for the irreversibility of time and, therefore, the infinite difference between alpha and omega, a difference measured by the full evolutionary process.

The environment, this phenomenon of which we are a part, is a small fragment of the immense cosmic genesis and as such is an ecotheological fragment; provided consciousness is capable of steering such genesis toward a seminal conclusion, omega. This time-loaded eschatology could be the ground on which science and theology might understand each other; and such understanding could beget new inconceivable fruits. In my perception, that is where a religious and scientific world must seek new artistic expressions as far away from the Garden of Eden as from any war-of-the worlds fantasies.

This model, where sainthood is the anticipation of divinity and where the esthetic occasion is a concrete if infinitesimal fragment of divinity, suggests that the Divine is the infinitely beautiful. Since the "proof is in the pudding", the eschatology can be truly validated only if the conclusion is reached. The end will justify and explain the process. The process is the suffered, experiential, step by step creation which makes up evolution. It is inescapable that such process is saturated with suffering and anguish, if for no other reason than we know that we do not know. But we must face the fact that we cannot know, since the future, that mesmerizing tornado made of naught, is the conclusive repository of the whys, the whats, and the hows.

The ultimate seed D.N.A. does not exist. It begs for creation. The chromosomes of such a seed are presently inconceivable as much or more than our genetic make-up is inconceivable to the intellection that a grain of sand can muster.

The theologian, free from idolatry, and the artist, free from hedonism, might turn to the environment, the ecotheological process, in awe and reverence. For in it, and by their perception (turning into intervention) they may find reasons for hope and exultation. The media, mass-energy and space-time, are neither generous nor understanding, but they may be part of a radiant message beyond anticipation. Therefore, the imperative for theologian and artist is humility and reverence toward the "tremendous mystery" of becoming. Thus a new environment is not simply a physical setting that responds more effectively through its resources to the problems of contemporary society (energy crisis, segregation, pollution, hunger, consumerism, etc.). It is also a perception of reality, alert to the shortcomings of animism and determinism, that might propose a more intense interplay between wo-man, the living world and the geophysical (and necessarily cosmic) reality. What this intensity will or ought to be, is for us at our best to determine, or at least to seek.

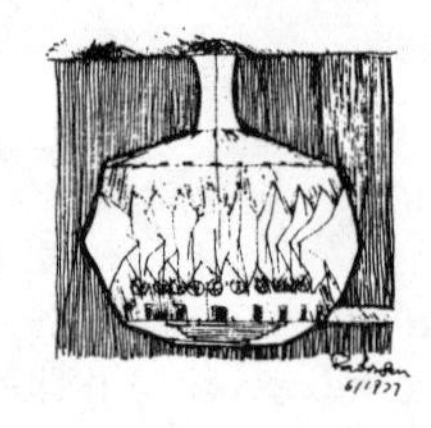

Chapter Eight

Animism, Technocracy, and the Urban Effect

Technocracy and animism are the nemesis of a reality that institutionalizes both materialism and revelation. Rampant technocracy is the obscurantism of a consumer society reduced to a processing plant: head and tail are indistinguishable; what goes in is what comes out; processed, yes; transfigured, no. Animism is a bottomless pit from which one can pull out any rabbit and throw in any idiosyncracy. You have something to explain. Pronto! The devas or their cousins willingly supply the explanation. You have something to justify. Pronto! The devas and bad relatives made you do it.

Given animism, the suffering, the worthy, and the graceful are defined as pale reflections or distortions of a perfection that preceded them and now hovers above. The reflection is not the substance and significance. For it is only an image in a mirror. It distorts the real substance. The substance is in the anima above and below, in and out, all around. At most, the reflection is a spin-off of a perfection itching for something. What could it be?. . . imperfection?

This "bricolage" of a God is a danger signal that forecasts deception. One has no great difficulty in acknowledging different interpretational modes and semantic uses that do not represent substantial disagreement. Therefore, devas and custodial angels can be called the mysteries of a history too long and complex to be rationalized; or inner motivations so pressing that they crystalize as colloquia with oneself. But the moment comes when interpretation and semantics do not neutralize the incompatibility of two positions. Then amicable dispute gives way to conflict between the two paradigms, and the conflict is real. Souls are thrown onto a battlefield where one "truth" is in deadly combat with another "truth". It is no limited or isolated conflict. Reality demands that the ripples originating at the confrontation's center agitate and propagate endlessly throughout immanence.

The nature and circumstances of the causa prima are now at the center of human concern, as they have been for millennia. Here they are an awesome question mark fatally coloring all human endeavours. Such concern manifests itself on all sorts of deceiving occasions. The more that intellection and knowledge deepen and grow, the more crucial is the influence of the *causa prima* on human behavior; not so much by way of interpretation as in its true nature.

This paradigmatic intolerance is the eye of the storm of cruelty and destruction in history. Proclaiming the truth and in its name putting the land to fire was the operating principle of Nazism, of the Crusades as it was for most attempts to bring salvation to infidels by conquering them.

We must acknowledge that whether there be two truths or a thousand truths, truth is a purely anticipatory notion. It is devoid of absolute truth, but full of longing for it. If we do not acknowledge this we will no doubt offer an endless number of sacrificial lambs to uphold or "incarnate" the truths that we have invented.

The principle of uncertainty is far more than a "scientific" fact. It is an ontological and eschatological inescapability. In the recognized absence of the truth, a non-idolatrous paradigm may be proposed. It will not be a definition of the truth but an inscription of its probability. What the mind may have to do then is to probe two aspects of probability: the aspect of feasibility within probability; and the aspect of desirability within probability.

Eschatology

No eschatological model can construct a normative paradigm based on truth without undermining its own premises. If there is the possibility of an end, that end is truth. If truth is the end, that which comes before is at best incomplete. Incomplete truth is not truth. Aggravating the untrueness of incompleteness is the fact that the greater the space-time distance from the eschatological end, the graver the risk of error entailed in presuming knowledge of a future end, that is not just shrouded in eons of becoming, but is nonexistent, since the eons themselves do not exist.

At this distance from the eschatological end not only are we ontologically truth-deficient but we also suffer from "embryological inexperience". We are hermetic and enigmatic genes intelligently scratching at the surface of an immense but dumb reality; scratching at outlines on the armour of a monster. And the monster is as yet incapable of hospitality or response; the poor thing is in the making. The eschatological project entails its taming and the consumption of its power for the sake of truth. The truth cannot coexist with it since that would mean that times are ripe and times are not ripe. Times are ripe when time is no more; consumed as it will then be in its struggle to transfigure the monster (of which time is one limb along with those other limbs of space, mass, and energy).

Once we acknowledge that truth does not exist, and its hypothetical place is occupied by processes (living organisms included) whose ultimate significance is yet to be created, then we are more willing to accept existential contradictions and ambiguities. By scaling down the dimensions of revelation to the point of redefining it as anticipation (or, as in less hopeful cases, spiritual sadomasochism) we may become more willing to consider that the miracle of miracles is not to "progress" from more into less but instead from less into more. To move from less into more means a creational process necessarily inclusive of the remote future of reality (the past) to be. The inner contradiction of animism is the pretense that a creational process is occurring but that, at the same time, fullness is a pre-established condition. This is not so much a mystery as it is a contradiction. If time is, fullness is not. If fullness is, time is not. In other words, fullness belongs to being, not to becoming. By putting fullness at the origin, we preempt becoming of any substance and leave it as a becoming world of appearances (whatever that might mean).

Consumption of time-space-mass-energy is the premise of a truth, that is and is all. The incompatibility of becoming and truth has anguished man throughout history and no religion or sect can escape the crushing weight of such reciprocal exclusion. To proclaim the truth is to lie to reality on two scores: (1) We know little of the past; (2) The future is unknowable not out of ignorance but out of non-existence. Therefore, truth is unknowable not only because of ignorance but also, more crucially, because that which will be is not now. This is the impossibility of truth as reality. *If there is a tomorrow, there is no truth (now). If there is truth, there is no tomorrow.* The creational process is the creation of truth. It cannot be assumed that creation is guided by truth since such guidance would preempt the purpose of process. The esthetic world illustrates the difference between the description of an act of creation and the substance of it. The two are incommensurable. The description true and total is the work of art with no possible substitutions. Truth is not the description, *a priori* or *a posteriori*. "Truth" is the objectified work.

Since the esthetogenesis of the cosmos is the ultimate work of art, there is no way that descriptive, anticipatory legislation, let alone finalized laws, can be established that will even vaguely define the nature and power of truth. Creation is not a process of being guided to the fountain of life. It is, instead, the creation of it. The effects cannot be defined or even described by the cause. To dread the enormous responsibility of creation must not end in a loss of anima so total as to preclude the possibility of mention; even if the probabilities of fulfillment are infinitesimal. On the other hand, to lull ourselves with facile notions of salvation, redemption, beatitude, peace, grace, or beauty is dishonest and flies in the face of history.

Justice

Of all deceptions, none is greater than stilling the conscience with the notion that the past is irretrievable and that we are therefore discharged of responsibility for inequity and suffering gone by. The "truth" is that the marrow of our bones is made of them. They are us, us "forever". At the end of forever they must be with IT, of IT, to glow in a *true* resurrection and gnosis. As I have said I call that ultimate condition the Omega Seed. Our seed child is utterly inconceivable; its genes are yet to be created so in no way can one prophesy its make-up. What one can conjecture is that for the Omega Seed, time-space-mass-energy are unacceptable since they stand for residual, segregative, uncommunicative, and unloving particulates. Its beingness is therefore mandatory since becoming implies that the creational changes are possible only when time beats, when space is measurable, and when mass-energy is the fury of separateness and incompleteness.

Reality is being accumulated. In fact, it is the accumlation of a debt that strikes one as irredeemable. It is the cost of making matter into spirit. It is the making of all deficiencies into fulfillment, into freedom, into peace, into a just society, etc. We are constructing our future on a raft adrift in a sea of blood and pain. And into this sea we continue to spill more blood while we go about our task that has practically no end. On such a landscape the ego is supreme and dementia prevails. But on a landscape where grace prevails, truth and justice create themselves.

Technocracy

The animistic paradigm refutes creation by asserting that there is more at the beginning than there is now. (It would seem that perfection plus imperfection equals less than perfection.) The technocratic paradigm refutes creation by asserting that the effects are built into the causes, that given time, discrimination, and horsesense the best of all worlds will be produced via the pursuit of happiness as a good management policy. Upon comparison, animism, and technocracy agree even if in animism causes are gathered in the *causa prima*; while for the technocrat-priest there exists a gradual growth of causes generating effects which in turn become causal for other effects.

The animist-priest usually dresses his originator in a benevolent or just garb. This fatherly or motherly figure that is depicted by the churches helps believers find motivations to improve themselves. There remains, however, an aura of predestination if not about the details, at least about the unavoidable triumph of godness, since godness is the stuff of the great one or ones, our Lords. The presence of the Lords preempts the urgency of bringing godness about because godness is. If godness is, I'll be damned if I don't pursue it, but my damnation is surrounded by pervading benevolence. If I am not my brother's keeper, God will see to it.

Respectability, the self-responsibility of the technocrat is split between two icons. One is the god of one's own religion who one may want to stay on good terms with just in case. The other is success. The "thing" must sell, three cheers to technology and to the (revised) Calvinistic god. In the technocrat the icon of success questions, but only amicably, the validity of predestination. The final pride for making it is part and parcel of the determinism of cause-effect. In fact, determinism is really the icon to be worshipped. The other icon, the denominational god, is a back-up, an insurance policy.

The Urban Effect

The Urban Effect is that fundamental phenomenon in which two or more particles of physical matter begin to interact in ways other than statistical and fatal. That is to say, in ways which are organic or living, and eventually the instinctive, self-conscious, mental, cultural, and spiritual ways. It is this Urban Effect around which the whole experiment of life is clustering itself.

In the Urban Effect as in biological mutation-adaptation, the cause-effect link is regularly, "systematically" ignored. There is no cause for mutation aside from the inherent fallibility of any organism in the pursuit of being absolutely true to its own genetic determinism; that is to say, to be unchangeable. Once mutation sets in, it becomes a causation effecting the development or the demise of the organism. The causation is much like the appearance of a mechanical defect in a self-perpetuating machine. But, while in biology chance is the "benevolent" agent for change, in mutation transcendence is the transfigurative power of change. The esthetic is "intelligent" mutation, where in place of chance as agent, one finds anguish. The reptile becoming the bird (scales into feathers, etc.) is not fully explained by a sequence of links alternating causes and effects. It is better symbolized by explosions (mutations) linked by bridges which stand for necessary selections-adaptations that the explosions (mutations) force a priori upon the living chain selectively progressing from reptile to bird. The mutations are not directed, they are haphazard and define the operational field within which adaptation and survival determine the fate of the organism (the species). Genes are hermetic machines. The organism has no authority or power over them. If one wants to introduce guidance, it has all the characteristics of the laws of probability and none of the father-mother benevolent sensitivity. This strongly suggests:

1) that father-mother do not have the numbers, that they are impotent; 2) that all the divinity invoked by religion is really an embryonic stirring capable of miraculous acts.

To clarify the difference between animism and the Urban Effect, I will return to the metaphors of the super jigsaw puzzle and the super rubber ball.

The Jigsaw Metaphor

In the *animistic model* reality is perceived as an immense mosaic. The image is unknown because the mosaic bits are scrambled. The scrambling infers a scrambler, a divinity or a divine power. The scrambling is not total: after a certain historical progression, consciousness has begun to see hints and outlines, vaguely defined images. Consciousness and a "love for truth" are the agents progressing through pain and exhilaration toward the unscrambling of the puzzle. The puzzle knows its own image all along but is somehow unwilling or incapable of formalizing it. Therefore, a cooperative action occurs between the original truth, the scrambled mosaic, and the "evolutionary" effort to figure out the image. In so doing, the cooperation becomes part of the effort. In fact, the unscrambler is one of the chips. It has to find its own position while positioning other chips (natural history, for instance). Depending on what religious creed one is wedded to, the puzzle metaphor offers emphasis that makes it fit the creed (see "Wo-Man—Space—Justice"). An extreme version is that of the mosaic already composed but hidden under a heavy covering of inert substance: ignorance. Theology guesses at the meaning of the mosaic. Science attempts to scrape away the covering of ignorance and uncover the true nature of the mosaic's image.

The Rubber Ball Metaphor

The Urban Effect (Omega Seed) involves perceiving reality as a rubber ball of very peculiar characteristics. The ball is initially infinitesimal but expanding. In expanding it generates the capacity to create "images." Some of those images become self-perpetuating and self-conscious and introduce their own will into the cosmic time-space stretching, while imagery proceeds on the surface of the ball (the present). Eventually, this herculean effort causes the sphere to halt its expansion and, in fact, initiate its contraction while the nature of its composition is continually altered (transcended) by a creational process that is now powerful and active. Within the shrinking ball, more and more, images crowd each other (matter becomes spirit) and the past is recollected by the present's pull on it. Eventually the rubber ball is transfigured into a pointlike super rubber ball (the Omega Seed). This metaphor may be more understandable if one imagines one strand of the many composing the ball to be that of a rubber band.

One can compare becoming to an elastic body springing from a point of origin (the Big Bang) because of an inner trust (the time-space-mass-energy urge). At a certain moment (the moment of consciousness), the tension developing in the elastic body is great enough to originate contractions within it. These contractions travel within the body in the direction of the head since the head by its trust draws upon itself the rest of the body. Therefore, after an expansion of the body (stretching of time-space), a phase of contraction appears (the consumption of time-space). It is a "body" launched into the future. It has passed the limits of the stretch, and has begun a process of self-contraction toward an imaginary station yet to be designed (Omega and the creational process). The rubber band expands and releases itself beyond the furthest point of its expansion.

It is the re-absorption of the past into the future by means of creational train of present after present that is becoming ever more enabled to recall and experience the past. The moment (the last moment) will come when all of the past finds itself stored, or better, restored, within the last station of the project: Omega. It will be a resurrection because of such restoration.

Therefore, reality is not defined by a "returning" from the future to the past, but by the moving of the past, all of it, to the last of all futures, the Omega. This agrees with physiological genesis where the gene is the presence of a selected past in a "future" organism defined by the ovum. But the genetic memory is only a pale anticipation of the full "memory", i.e., re-enactment of the Omega Seed.

Technocracy is less than interested in either model. But one could compare progress to a progression of events somehow fitting a lead ball model. The rigidity of the material causes the substance of the ball to crack. It is a tired inelastic, distressed ball, a plethora of parochial segments traveling on the path of feasibility and obsolescence.

Omega Seed is the past stored in the future.

But given that the future is non-existent, the past is unreachable. It is secretive, unknowable, and unexplainable in absolute terms. The future holds the key to the riddle not so much because knowledge will flourish by its creation, but because creation itself, the gene of evolution, is the key. *In creation is the future that "stores" the past; in revelation is the past that "stores" the future.* In the second mode there is no way for novelty to exist because the first mover, the pristine past, the *causa prima*, pulls the strings and evaluates, judges, punishes, rewards. Truth is the first mover, unbreakable, unperfectable. In the Omega Seed the "last moment" is the judgment, the last judgment. The acknowledging, the conclusive, the totalizing, the truth defining itself to itself. The judge will not be a god but reality itself; not judging really, but being in fullness, in grace, in knowledge, in beauty, in equity. At that stage, the conclusive stage, there is the achievement of genesis. The genesis concluded of one's own seed, carrier of all of itself. Love for the past is in the creation of the future because love for the past entails the passion for its "redemption" and its redemption is not in the creation of a better future per se, *but in the creation of that future which will bring about resurrection and the universal flood of equity which is one manifestation of it.* The grace of the future, even the greatest conceivable grace, cannot cancel the inequities of the past because the past is made of specific occasions of sufferance. Only if those occasions of sufferance are made alive again can the past and all of reality enter the realm of equity. It is not simply a question of being worthy of the sufferance that came before us and made us. It is the question of making all past occasions fully alive in the light of equity and grace. There is then the most important imperative of absorbing the past, collapsing it into the timeless present of the conclusive state of the Omega Seed.

The thesis is simple: the cosmos is involved in an attempt to generate its own seed. Cosmogenesis is the creation of its own genetic matrix. The super egg is not "dropped" by a pre-existing cosmic superbird. It is instead the receptacle of all that the cosmos has been while creating itself in the future, until the consumption of all futures. The cosmic seed will then be all that there is, and what there is, is all that there has ever become.

Justice is retroactive or it is pseudo-justice. Justice can be retroactive only if cosmogenesis becomes esthetogenesis, the "thing" itself (not a synopsis or a description of it), the generation of grace and beauty to a degree which coinvolves the entire cosmic reality. No matter how we manipulate the models of reality, unless that which has gone by, and all of reality will eventually have gone by, can be retrieved, re-presented in all its details, the immense sufferance of evolution (the food chain perceived physically and metaphysically) will remain unredeemed. One cannot undo the past. Even if the last instant of the process would be able to explain the components of it (the single occasions stretching in past time and space) there would be no justice; therefore, no grace. To do justice will require the perception and the enjoyment of justice by all occasions whose summation is the (hypothetical) last instant (The Omega Seed). Short of this universal guilt will remain a rampaging monster, a devourer of grace.

Beware of religions and its false paradise. Since the devil is ultimately entropy, there will be no kingdom of god until entropy fades to an impotent and forgotten notion. It is a cosmic battle that science defines as uneven and fatally determined. For today's science, the laws of physics doom the spirit to a temporary or fringe phenomenon. Justice and grace, the esthetogenesis of the cosmos are illusory, the hubris spoken by an irrelevant event (life). On the contrary, recognizing the real as the limited, one wants to make a hypothesis about the unlimited, the unreal, because then from the light of "something to come" (the unlimited) the limited may find guidance and ultimately its "remission of anguish".

If a hypothesis of divinity is legitimate, indeed imperative, the probability of its creation is next to infinitesimal but that hope is all we have with which to support ourselves. Without this outside chance of divinization, the savage living processes bare the brutishness of the cosmic structure. In doubting Camus' pessimism one grafts such doubt onto an infinitesimally weak stem. But doubt one must. For if the notion of divinity did not exist, it ought to be invented. Once invented it has to be "fitted" to reality. It has been misfitted. Reality is too frightful if it is not bridled by an intelligence, a loving intellect. To doubt the existence of the driver and the primer, be it love, nirvana, Jehovah, Allah, etc., is to bear an unbearable responsibility. We are the bearers of a cosmic responsibility and we respond with a cosmic cowardice that is *the* original sin. We pretend that our gods are real, we pretend that grace exists. We pretend that the pursuit of happiness is the gift of existence. Reality catches up with us, mandatorily. Individually and collectively we are delivered and deliver sufferance, torture, death by and among ourselves; while suffering all the afflictions nature in its "wisdom" rains upon us. But we go on praising the Lord. Be *it* damned and its insanity with it. For if I am wrong and God is, then my invective is too puny to matter (and I'll be damned). But if I am right and God is not, then my invective is far from being adequate to the enormity of the fraud and the sufferance it causes.

Is Omega Seed the grim substitute for the utopia of happiness? The greatest lust for life is in the nature of Omega Seed. For it, the fleeting moment never ends. The make-up of reality is the totality of all fleeting moments. Those fleeting moments are present after present, each taken as a timeless slice of reality. They are moment after moment. They are the cosmos. To lust for the Omega Seed is to live each fleeting moment of the cosmos in the durational infinity of "the last moment", as the resurrection of all the fleeting moments that had become aware of each other and are now the occasion for all occasions to coexist in and rejoice with one another.

Chapter Nine

Coevolution and the Seed of the Cosmos

Because of the birth of consciousness within a quite possibly meaningless universe, and because consciousness is mindfull of values, reality might now be seeking and pursuing meaning.

My fundamental hypothesis is that reality, "equipped" with consciousness, is attempting to create its own semen. This reverses conventional theological thinking. What religion places at the origin —the seed, the source of divine power, the fountainhead–is to be seen at the end instead. Cosmogenesis, as the word says, is the creation of genes, the ultimate genes of the cosmos. In this case, such process involves the total consumption of the media available, the physical cosmos itself. Then, at the final juncture of consummation, the cosmos would replay itself completely (resurrection) as any good seed replays itself through the organism it "contains".

I am not peddling the truth, I am presenting only an hypothesis. Why, and why one such as this. I would follow this line of reasoning: if it is a fact that equity, justice, truth, beauty, and grace are the real longing of conscious life, then this eschatological hypothesis might be the only one true to such longing. And if this were to be proven, then the imperative of consciousness clearly would be the full and unwavering dedication to the development of this ultimate seed, this divine, pure, total. . .unimaginable occasion.

According to this hypothesis, the truth does not exist, it is in the process of being created. The same is true of justice, perfection, beauty, and grace. The embryonic stage of all these at present explain the existence of injustice, squalor, suffering, disgrace, and ugliness.

The human existential dilemma is the discrepancy between anticipation and the context in which we live and act. We dream of perfection and we get imperfection; we dream of justice and we experience inequity: we dream of grace and we live in squalor...

The amount of truth represented by the past is known in very tiny parcels. Even more distressing, the bulk of truth does not yet exist because it is stored in that which does not exist, the future. Therefore, we know that we do not know (the past) and we also know that we cannot know (the future). And so, since the resolution of the cosmic puzzle is in the remoteness of a distant and far-from-granted future, anguish is in the very nature of conscious life. In the absurd case that we should have access to Knowledge, capital K, creation would stop, life would be extinguished, time (change) would end. This is a worry we need not waste time on.

But truth might come about through the self-creational process; therefore, time will be consumed. When this cosmogenesis is concluded, the cosmos will have achieved the creation of its own seed; because time will be no more, it will be the seed of resurrection. This is the total recall of the evolutionary process of cosmogenesis. The consequence of this conclusion will be justice, beauty, grace in full. Because of the retroactive nature of the Omega Seed (time-less), every occasion will have found its absolute fitness within the totality as well as the radiant and living knowledge of its own value. This is a skeletal explanation of my eschatological hypothesis.

In building the small town of Arcosanti in central Arizona we are trying to be congruent with that hypothesis so as to add an infinitesimal brick to the construction-creation of the ultimate seed. In other words, we are pursuing the Urban Effect because it is the most comprehensive process by which truth, justice, congruence, and beauty can be sought. The reason is simple: cosmogenesis is a process of complexification profound beyond the capacity of the most fervid imagination. The urban mode is the prototypical structure for fostering such a process.

The Urban Effect is not a human invention. It is the very common mean by which life itself goes about existing, persevering, reproducing, and evolving. Evolution is the Urban Effect imposed upon non-life by life. An imposition by the exception, life, upon the rule, non-life.

Based on the Urban Effect, *and congruent with its nature,* we are trying to establish a better environmental coherence, a better use of the resources of the sun. We are trying to reduce isolation, alienation, segregation, greed, intolerance, and prejudice. We are battling the mediocrity of a society hell-bent on consumerism.

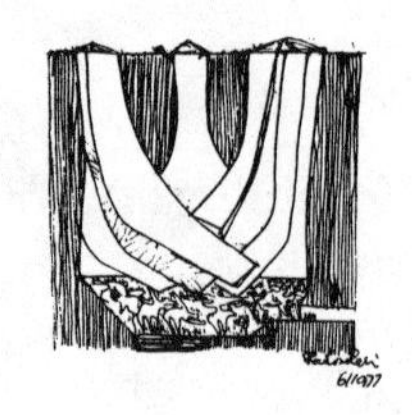

Chapter Ten

The Energy Dilemma

We need more than our current philosophy or ethic of life. We need not only to state a new philosophy and ethic but pursue them. For in our present situation the term "current" in the energy crisis should be taken out and "lasting" put in its place. With this as a given proposition, most of the present arguments pro and con on the energy situation should be seen as part of some theater of the absurd.

Consider the case of oil.

Imagine if you can, and necessarily in an ever-so-pale way, the 3000 or so millions of years (1000 times longer than the existence of the human species) and all the layers of livings and dyings and all the *complex* interferences and interactions that have eventually resulted in oil deposits; now such an object of wo-man's greedy and transfixed attention. The history of oil is a sustained, eon-lasting drama that has seen "simple compounds" from the mineral world, become *complex* organic compounds full of packed energy. That story is a small part of an even more astonishing story; the same mineral stuff becomes living, acting, self-perpetuating, emoting, thinking, dreaming stuff by means of a similar process of complexification. Now what do we do with this precious fluid? Mainly we create controlled explosion processes (burn it up) to propel something. Mainly we propel our bodies and our hardware. But to what purpose? Mainly we do this to reconnect (and feed) that which we have carefully disjoined, separated (and starved) by means of a series of discrete and modest choices that have resulted in the most wasteful, sterile, segregational landscape ever invented. I am sure that our visiting Martians would be groping for coherence seeing us gasping for clean air and ethical nexus. For both clean air and ethical nexus are becoming scarce resources, "endangered Species," if you will. And all along we propose our Western way as *the* way.

What I am about to say I believe to be in full accordance with the teaching of evolution and history. It also is said under the proddings of a present in search of a future. First I propose that the only way we can begin to justify our undoing of oil into those *simplified* components that we "refine" from it, is if we can demonstrate that a more *complex*, i.e. more lively equitable social-cultural, civi-lization is actually proceeding from this process. If it turns out not to be so, then we are surely committing the greatest sin of all, the "sin of entropy," the undoing of what the past earth has developed: We would be reversing an earth process that brings spirit out of matter, sensitivity out of coarseness, compassion out of indifference. . .

It is a sure bet that when we have (for all practical purposes) exhausted our oil deposits, we will realize too late that our squandering has forever deprived us of our most versatile of all compounds. For oil is a compound capable of performing miracles for us of a far greater magnitude than to serve our idiosyncratic notion of autonomy, freedom, happiness, etc. Fertilizers, chemicals, plastics, and the powering of our industries in addition to rubber wheel mobility do not (even remotely) do justice to the treasures that oil, gas and coal are.

To pretend a real concern for and a true understanding of the energy problem without showing a perception of the energy squandering that is built into our consumer-environment performance (the structure of our habitat villages, towns, cities) is pure deception. If then we think that we are worthier because individuals in our society consume as much as 30 times what an Indian does, then we our society should be prepared to realize the gloomiest of technocratic prophecies: a problem of exploding means and withering ends.

Today as in other momentous occasions of history, its actors are unable to put their fingers on the true meaning of their actions. This is true even when their actions are of such manipulative-transformative scale as to be the central cause of momentous change. It should by now be clear to anyone that the energy crisis is but a misnomer for that explosion of greed hidden within in the consumerist proposition. Tone down consumerism and greed (human diaspora) and the energy crisis fades away. Stay with consumerism-sprawl and the energy crisis will starve our world population of over 4 billion people.

Globally, the answer is ecotheologic. It seems that it does not make a great deal of difference how we go about satisfying our (Western) gargantuan lust for energy. As long as we are unwilling to confess to the pathology of unbridled greed and environmental extravagance that has brought about this situation then we will not change a syndrome that is a self-fulfilling prophetic catastrophe.

What are the ethics and the equity of a society unwilling to fast a little, abstain from meat once a week, have a somewhat more modest vacation budget, have one less room added to the house or have one less appliance etc., etc. when about one-third of wo-mankind (over 1000 million people) are undernourished and poor. The paradox is that many Americans try unsuccessfully for the most part to cut down on eating by pouring money into expensive dietary frauds. *We are in a major crisis of the spirit and as a consequence we have an energy crisis or weakness of the spirit within us. We must begin a radical and strong cleansing of the present system and follow this by a new policy of preventive medicine so as not to return again to an overstuffed organism. There is no limit to our "need" if we do not limit our "greed".*

Therefore:

Proposition No. 1.

The recognition of frugality as a virtue and its pursuit at all levels.

Proposition No. 2.

The increased use of "income" energy: Sun, wind, tides, etc.

Proposition No. 3.

The reduced use of "capital" energies: Oil, gas, coal, etc.

Proposition No. 4

The overcautious and equitable introduction of exotic technologies of fission, breeder, fusion, etc.

Proposition No. 5.

And most fundamental, the progressively staged transformation of our habitat so that in the same proces we cope with waste, pollution, depletion of resources, environmental destruction, encroachment upon nature, segregation, cultural and social degradation, inequity, mediocrity and ugliness.

In a reinforced, spiritual, civi-lizing nexus the so-called energy crisis will lose its grip and substance. A new freedom can become available to a more disciplined inner-oriented society composed of self-disciplined, self-responsible (altruistic?) people.

Going solar only is pure escapism and a retreat into self-righteousness. Often Solar people feel down deep so many times smarter than those thick-skulled nukes. And the nukes reciprocate with gusto and reality gets lost in a loss of anima on the part of both parties. Love your enemy and care for your adversary become irrelevant. There is no energy crisis. It is a spiritual crisis.

Thus I reject the existence of the energy crisis per se. Do I then find myself a bedfellow of the technocrat who proclaims the energy crisis to be a political fabrication? No. I propose that the energy crisis is the result of technocratic "fabrication". It is the unavoidable consequence of a materialistic option in a society equipped with a formidable technological know-how and an unchecked propensity for a quiet but pervasive greed.

If going solar means becoming illuminated, then it promises the means for a more equitable and lively society. But only if it is seen as a means to dig ourselves out of our self-inflicted consumerist entombment. Otherwise going solar might just mean smarter gadgetry packaged by the intolerant and self-righteous and with a spice of obscurantism. It may just be tied to kilowatt hour curtailment when the true social-cultural nexus is to be found in a greater complexity of social context. In these terms, going natural may bring us face to face with a dreadful cultural-scientific technological collapse; one in which the institutions of our civilization become extinct and with their place being taken over by some agrarian-arcadian mono-culture that inherits the suburban culture by default.

Also going solar may have little to do with massive farming of electricity from sunlight on the surface of the earth. It requires the raw materials that must be used in the process of producing the necessary hardware. It involves possibly greater problems because of the handling of large amounts of these raw materials and especially rare elements (freon, etc.) that are needed for the running of the plants. As well, there would be an ecological impact resultant from transforming waste areas of ground into sun collectors. With all of this a technology of outer space collection of energy might turn out to be more attractive.

Thus strangely, going solar might ultimately mean going along with the same ecological situation as we find ourselves in. But if this is not congruent with our specific social-cultural-spiritual nature, then it will support the further demeaning of our true humaneness. To become congruent is to find the true hinge (interface) between wo-man and nature. *It goes without saying that this means to rediscover the authentic city*, to find the true meaning of the urban effect and in so doing to define the best match between nature's radiance—the sun, and wo-man's radiance of the spirit.

To put it into a historical perspective, it might be reasonable to propose that some of the most distant architectural-urban solutions developed throughout history might have to be perceived not so much as the blueprints for our very different times but as methodological propositions that might be useful if translated into a modern context. In addition, we have much new scientific knowledge, new technologies and new materials (i.e. the transparent-translucent membrane).

The Arcosanti project is an idea and a process that attempts to reflect a congruence of nature-wo-man within a historical continuum, and with a the scientific-technological ("passive") presence:

1. It does not reject the existing technologies of energy production but tries to define a more responsible and less intensive use of the resources that they make available.

2. It does not seek energy self-sufficiency not only because this is impossible practically within a fully operational lively community (urban effect), but more basically because it is ontological nonsense.
3. It forcefully rejects the flat grid layout of the existing communities as most wasteful in resources and in human terms. In doing so it demands the exclusion of the automobile from the habitat and the reinstatement of the two-legged citizen as its sovereign presence.
4. It advocates greater integration and proximity of the diverse activities of people, along with a parallel desegregation of institutions and facilities, among them food production processing, marketing, learning, performing and enjoying etc.
5. It does not advocate the sensitization to the sun of every single structure (ultimately incongruous) but it seeks to bring the whole urban landscape into the presence and action of the sun according to its seasonal character.
6. It advocates the self-containment of the urban-scape for the sake of frugality; instantaneous access to the open land; the impact of the contrast city-scape countryside; the pedestrian-bicycle mode of the community; the desegregation of the social-cultural, economic, productive, commercial, context; the dismissal of the empty street-precinct syndrome and the danger of alienation, vandalism and unlawfulness which it encourages.

The Arcosanti project is intended as an urban laboratory. It seeks to become a trigger for a directed search for those urban alternatives now ignored or dismissed. A loss of anima is suggested to me by a society that unfortunately endorses the democratic pretenses of consumerism. For it certainly slams the door tight against the mounting pressure for equity that is building up within each of us (in our better moments, at least). The affliction of contemporary society is a penury of spirit which reflects itself in a scarcity of (physical) energy. This "economy of scarcity" brings to the surface the least commendable of wo-man's character: Fear, greed, intolerance.

The energy shortage should have been the trigger and occasion for revisiting architecture. It may be too early to give a very definitive assessment of what has been triggered. But it is not too soon to say that intellectual and moral cowardice is presiding over the architecture community. The sad thing is that we are unable to perceive the entropy of our present position.

It should also come as a most welcome challenge that the two questions of energy and the city are so much in the mind of planners, economists, politicians, and social scientists, not to mention technocrats.

Unfortunately, it seems that each of them seeks optimization of their own separate field and as a consequence they do not communicate with each other. If they do listen to each other they only hear what they want to hear. Only by optimizing compromise and ambiguity will be found a human response to existence (becoming); otherwise, optimism can be the nemesis of wholeness.

While the "sun god" fries eggs on our technological contraptions, his sons and daughters seem incapable of identifying and recognizing the religious, ecotheological challenge of the energy-sun-environment-spirit proposition. By being blind to the true nature of the proposition, we are forever bogged down in a technological no wo-man's land taken up for the most part by corporate and appropriate technologies.

If small thinking is often beautiful, fragmented thinking is never so. And our urban landscape is plagued by fragmented thinking. Fragmented thinking is not greed, but certainly greed is fragmented thinking. In fact, the two go together both to support and rationalize the present urban catastrophe.

It might be argued that fragmented thinking, like democratic procedures, may be the best way of avoiding catastrophe. In order to classify something as a catastrophe one must depend on the numbers. A pot broken into a few fragments is a broken but repairable pot. A pot shattered into smithereens is broken beyond repair. It has lost that which makes it a pot. The shopping center is the repository and icon of shattered wo-man. It is an accumulative result of many innocent actions. All of the millions of articles on display can be individually explained and justified. But the consequences of their collective existence is less than edifying. The suburban shopping center is the consumerist society at its best, and the urban landscape is one of its expressions. It is greed and fragmentation, with an extraordinary assertion of human scale appended to it.

It is easy to measure or perceive a sudden catastrophe. But it is far more difficult to do so when we become part of one. We must rethink the catastrophe we have grown into (and produced), and only then will we rethink architecture, no sooner.

A society incapable of a quality of boldness of the spirit is a listless society regardless of the momumental quality of its gross national product.

The frugality and stringency essential to a well-balanced organism has been forced out of the urban organism. Kilowatt hours and horsepower, the technocratic emblems of happiness, crowd together in direct proportion to the listlessness of a system: "Do not make do with one, what you can do with two. . .or ten etc."

The concept of the Urban Effect, that of the saturation of complexity and the saturation of ambiguity in all living events, is an endorsement of a process that sees an original media (mass-energy, space-time) making itself into sensitivity and intellection. It is an energy process, a metamorphosis, that in its most subtle manifestations takes the urban form. Complexity can only develop within miniaturized and miniaturizing containers. By the existence of complexity-miniaturization, physical time becomes duration. The Urban Effect is metamorphosis, hopefully a metamorphosis freed from cyclicity. It expresses itself in tangible and intangible ways. But the existing urban grid has less and less to express and is not capable of good performance.

Unless the urban context clearly expresses and effectively implements the complexity-miniaturization-duration paradigm, its core will die. There is one single technological device that assures the coming of urban death. It is the automobile. In terms of equity, coherence, and resilience, death has already come. Esthetically, bits and pieces have escaped doom but are more and more immersed in the death sea of the supermediocre. Irony is sought rather than transcendence. And the esthetic that is there has primarily a soul shaking, bad economic and materialistic underpinning. Such an extravagant, wasteful, and incoherent city is an esthetic nemesis.

The energy question is the loss of anima question as I have said; therefore, dealing with it as we go, we are eons away from the understanding its true nature and thus find a way out of the dilemma. Sooner or later the whole cosmos will open its reservoir of energy to wo-man. But it is possible that by then that wo-man might be de-energized, worthless. In this light, retrofitting with solar panels and all that this symbolizes might well be insult added to injury.

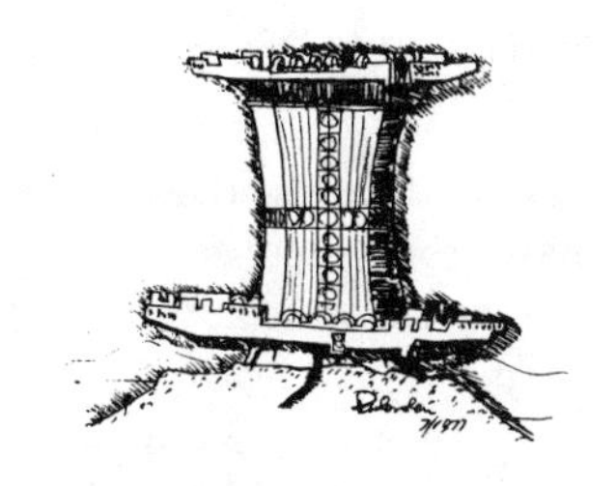

Chapter Eleven

Organic Structure and the Potential of the New Community

Within the continuum of reality, there are necessarily sharp hierarchical distinctions. But egalitarian minds in their longing for equity often come up with invective scenarios against hierarchical discrimination and throw everything indiscriminately in the organic pot. Yet, in absence of hierarchy, a mushroom can be grafted onto a spinal cord or a stone in place of a heart.

Such misplaced equity rubs against the image of that which is supposed to be organic, for what is truly organic is the epitomy of hierarchy. A misguided view says that the organic is that which merely has the *appearance* of a "living organism" or, minimally that *it brings to mind* a living organism since such forms are the ways of nature. They are, it goes on say, *of* the *natural order* and in homage to such a superior order man's artifacts must *look like they are* organic. They are surrogates of the organic.

Not so, I say. The true organic is that which fosters occasions that are intensely selective, very discriminative, produce and transmit information of an intense character, solicits generous responses, and hastens an emotional and spiritual tone from participants. The true organic must have a self-transcendent capacity. So much for the distinction between what is a true organic form and what is purely a fake biological shape.

To offer something will not shortchange wo-man, we need to qualify the question of natural order. We need to do this even if it may turn out that natural order disqualifies itself. Let's see where this takes us. The notion of *New Community* must make sense transnationally. I propose that transnational meaning is impossible unless wo-man's consciousness points to a transrational meaning. This has been proven again and again in the rise and fall of civilizations. Marxism makes the argument that precisely because of an interjection of transcendental values civilizations have failed to produce *new communities that fit the natural order.*

My counter argument is that the transcendental values of which they speak have been of an idolatrous nature and, therefore, could not but produce negative consequences. Intolerance, self-deception, bigotry, gullibility and simplemindedness are such festering forms of idolatry. And I denounce them as direct consequences of an animistic view of reality. As I have said earlier, *The natural order and its offspring, the revelational order, are both animistic.*

There seem to be, as was said, two main emphases in the assessment of reality:

The natural order, a fatal assessment

The revelational order, a salvational assessment

Both are unsatisfactory.

The natural order model is unsatisfactory inasmuch as it tends to define time as circular or reversible; and tends to dismiss causality terms (with effects being consequences of causes) and thus along with it the responsibility. Indeed, within a reality where time is reversible, consciousness and, and even more conscience are irrelevant.

When Hinduism, Buddhism, and even more so, Taoism and Quantum Field Physics accept the reversibility of time they do not show even a desire to transcend much less offer a way to such reality. Therefore, and not withstanding claims to the contrary, they leave the reality of the investigator outside the investigated. In their acceptance of the reversibility of time, they forfeit all future time and thus any possibility of change.

The natural order is quite capable of explaining what is at the foundation of reality and of suggesting the tendencies that such reality might have.

The Tao and Quantum Field Physics are being used to bridge science and religion. But if a hidden postulate of both Tao and Quantum Field Physics is that the mind does not qualify, does not pass the test of reality, then what? In Tao, the mental reality is expedient. In Quantum Field Physics it is not quantifiable. One should not be concerned though that the disqualifying models, Tao and Quantum Field Physics, are the product of the disqualified phenomenon, the Mind.

Contrary to Tao, and perhaps to Quantum Field Physics, where the void is all there is, in revelation-salvation there is a divine originator, a *causa prima*, the Father-Mother, perfect and loving. The revelational-salvational model, inasmuch as it postulates a moment of pure perfection and plentitude at the beginning of time, also renders time illusory, ineffectual and irrelevant. Beyond the first plenitude of Jehovah, God, Allah, nothing is possible that amounts to more than, or as much as, plenitude itself. From this is the ultimate irrelevance of time and life.

Our gods, whether they be static or in agitation, are idols. They are idols because they are our inventions. The inventing per se is a glorious and forceful event, but the misplacement of meaning that follows, that is, giving to invention the arcane shroud of absolute revelation, is another story. We invent our ideals under the duress of the human condition. Under the prey of dread and fear, possessed by longings and passions, we name our ideals in different ways. We give them different attributes by stringent but a limited anthropomorphic logic. By putting a capital T before truth at our origin, we freeze the whole dynamic of our creational process.

Now, generations later and more knowledgeable about the ways of the world, we must transform reality in such a way as to make those inventions into a new reality; because our absolute gods do not exist for us as yet and they will not exist for eons to come. That means that the *natural order* (i.e. Tao-Quantum Field Physics) must be upgraded into the *surnatural order* that has been anticipated by these same revelational-salvational models. Willy-nilly we are engaged in the creation of such surnatural order, but we go about the task rather timidly and often cowardly.

There are strong signs now that wo-man may move once again and openly into the obscurantism of Holy Wars (not so much the natural order religions, but the revelational religions) and the recrudescence of cultism adds fuel to this fire. I maintain that the fundamental determinism of an animistic universe, be it Tao or God, ultimately gives us the most fertile hotbed for idolatry.

On one side the anthropomorphic models of Judaism, Christianity, Islam have and will cause virulent confrontations as ways of to upholding at various degrees of contamination, the uniqueness and trueness of their gods. Holy Wars are for the benefit of infidels. By blood and fire, their souls will be given reentry into the benevolence and everlasting bliss of the true god.

On the other side, the Tao, Brahma, and Nirvana models are strangely animating a reality that Quantum and relativistic physics are discovering to be at the foundation of the cosmos. With this animation that science defines as mass-energy "acting" in a space-time continuum, we are sold the notion that reality is a dynamic process. A closer look may reveal a notion which is quite different, a once-and-for-all given and determined reality, endlessly surveying and resurveying itself.

The given (Tao, Brahma, Nirvana, etc.), no matter how mobile, is, by its own definition, utterly rigid, fatal, unchangeable, indifferent. If and when the music is registered on a record, the gesticulation of a performer is beside the point. The music is in the record, Tao. The gesticulation is monkey business, life.

In its unending, cyclical dynamic, Tao defines a fatal reality frozen in its substance, random and deceptive in its appearances. In such a given, unalterable wheel of fate, consciousness and conscience cannot act because reality is indifferent to both of them, they do not matter. Such totally achieved model is available for any totalitarian drive because anything goes when nothing is really the causation of anything.

A universe in stupor is offered to wo-man by gurus of both sexes; a universe in which to dissolve oneself, where we can find ourselves as everything, everywhere, for all times. The problem with such a universe is that it looks more and more like the quantum relativistic reality of science. But isn't that the same reality that we, the living stuff, via the anticipatory models of monotheism, want to transfigure, recreate, consume into divinity? A true, loving, compassionate, knowing, divinity? A divinity-in-the-making? A divinity which is not a pale reflection of the Father nor a manifestation of Tao, but instead an embryo which already makes them both dull, brittle, inadequate instruments that the consciousness and the conscience of wo-man is using in the quest for equity, knowledge, grace?

What does exist within the gods and paradises of the animistic universe is a tidal wave of evolution and a food chain that, so to speak, is doing its work. All angels and vedas, all spirits and holy ghosts are made whimsical vis-à-vis the immense and harsh phenomenon of this life generating, preserving, creating process.

This protracted and incorruptible approach to life that is constantly creating spirit out of matter is not possible under the shadow of or in the light of a pre-existing anthropomorphic god; nor is this even considered in the fatal landscape of the eastern void. Such will only proceed by way of a process of consuming all the resources that such a void is made of, and according to patterns only obscurely anticipated by the religious inventiveness of the monotheistic prophets.

The process of consuming of an immense universe would, in the end, incorporate all of it, all of its evolutionary turmoil and all of its occasions. It would exclude nothing and it would by then be all-comprehensive in the glow of a knowledge that is total, a love that is total, a grace that is total, a reality that is total, a justice that is total. Within a more limited range, the monotheistic religions propose a similar occasion and call it resurrection.

Well, this is not the *natural order* as it is not a reflection or a derivative of an eternal Tao or God. It is instead, and in opposition, a most improbable, problematic, protracted, remote and suffered embodiment of all conceivable hypotheses. It demands most from reality because it wants to make the most of it.

This hypothesis cannot be idolatrous. In fact, it is the hypothesis that can transform idolatry into religiousness on at least three counts:

1. It recognizes the non-existence of its aim because it is an aim which is in the future and idolatry cannot be about that which is not.

2. It recognizes the possibility of failure and, consequently, the possible tragic waste of the whole evolutionary process.

3. It reverentially acknowledges the fundamental anguish of a divinity in the making, which is the whole of life and us within it.

The Bhagavad Gita, the Bible, the Koran, and all the sacred texts need defrocking. They must be changed from being revelatory documents spoken by the gods and transformed into culture-defined, time-bound, historical documents spoken by men inspired by fallible longings and anticipations.

The *organic structure* in the direction of the past is the structure of the *natural order*, Tao.

The *organic structure* in the direction of the future is the structure of the *surnatural order*, the order anticipated by the monotheistic religions.

In the direction of the past we see a reality that unwinds and simplifies itself, because in an evolution running in reverse, the mental, at first, and then the organic are progressively weakening and then disappearing, leaving the inorganic substratum to perform its own withdrawal (the Big Bang in reverse).

In the direction of the future we see a reality that complexifies itself, where, on the inorganic substratum, first appears the organic, then the mental and then, in ever-growing waves, the spiritual.

Such a process of complexification is powerful and has a staying power now measured on earth at almost 4000 milllion years. Therefore it is not extravagant to anticipate many more billions of years of complexification. At the end of the tunnel of complexification may well be that light which the revelationist-anticipatory religious models define as divinity.

Focusing now on the theme, *Organic Structure: The Potential of the New Community*, I am proposing that the associative, productive, cultural, scientific, esthetic, compassionate, religious phenomenon that we call towns and cities are the crucibles of such complexifying processes at the collective level of consciousness. I call such a phenomenon the Urban Effect. I am proposing (in congruence with this) that the Urban Effect is indeed that creational thrust that saturates the evolutionary process; because the drive that is carrying with it the urban phenomenon is also driving any and all organisms and association of organisms. Such a drive originates from them all. There are strong indications that consciousness is taking upon itself the steering of such a drive, and responsibility for the good and the evil that it generates (metaevolution).

Finally, I am proposing that new communities will be really new, and will be really dedicated to wo-man's well being, only if they are concrete, performing occasions for the Urban Effect to intensify itself and to create a transnatural order. Any return to the natural order may, therefore, be out of order because it proposes the rejection and halting of Urban Effect evolution.

As I have said Arcosanti, our attempt to construct an urban laboratory in Central Arizona, is conceived within the Urban Effect frame of reference. In such a frame it is a rejection of idolatry, intolerance, fanaticism, cultism and has no use whatsoever for the notions of the chosen people, the perfect society, the Garden of Eden, utopia, paradise and so on. All those notions are incompatible with the hypothesis proposing that reality is a process of true creation; because for such hypothesis, truth is in the making, that is, it is yet to come.

Chapter Twelve

More on the Urban Effect

Twenty years ago I proposed that physical, emotional, racial and functional segregation were a greater evil than waste, pollution and environmental destruction. I think that I was on target then and that I am now. It is quite possible that in our neurosis about gasoline shortages we do not see the energy crisis as it is, a blessing in disguise.

The energy crisis will force us to do what we should have chosen to do long ago: put some yeast in our city-pancakes so as to thicken the life they can sustain. Such "levitation" will hasten, to a point, what I have referred to as *the Urban Effect,* a condition in which the vicarious, cooperative, inventive, social, cultural, creative propensities of wo-man can have a more favorable milieu in which to flourish and at a fraction of the price we now reluctantly pay or refuse to pay.

Since the Urban Effect is at the core of life, all life, at our respectable age of 3,500 million years, the Urban Effect might well become an eschatological stress. The Urban Effect, in any case, is no abstraction. It is not the Almighty; it is not Super-Robot; it is not Big Brother; it is not the Masses. It is the personal, the you and I engaged in that amount of self and collective fulfillment permitted by the context and constraints of reality, the living and self-creating, within the natural order (TAO) of quantum-relativistic physics.

The need for myth has been reconsidered recently. We have built the technocratic myth with the automobile as the emblematic image and sacred icon. Together with irrefutable advantages to woman, we are also getting disastrous conditions and an inherent fragility; for instance, the real possibility of partial or critical paralysis of our society. But perhaps we do not need myth. We have the universe. In it we have life's evolution running on the food chain storm and we, humans, are on the devastatingly and exhilarating cutting edge of it. In fact, the storm takes us, willy nilly, with itself. It is by now a tidal wave. We are its coauthors and we are thus coresponsible for it.

The roaring evolutionary tide is disciplined by the resources available (brain power included). The resources available are organized and structured by the so-called laws of nature, the electromagnetic, nuclear, gravitational rules. We abide by the rules or we are done in by their very existence. Gravity and thermodynamics are routinely, moment after moment, giving, taking, prodding, padding, smacking us, reminding us of the limits of our extravagances.

Then at the top of the physical substructure of resources is the food chain disciplinarian. But this is not a benevolent master. For instance, we butcher hundreds of thousands of animals day in and day out. Then, in a perverted sublimination of the food chain imperative we butcher each other determinedly or, out of tedium, on a quasi-metronomic routine.

To make coherence, equity, dignity and grace slightly less than a dream, we need to put a greater self-responsibility in our actions. Believe it or not, the Urban Effect is the comprehensive mode for that. It is the greatest transformer of dullness, of the resources of mass-energy-time-space, into sensitivity, mind, conscience, anticipation, by way of a compassionate inconsistency with the food chain's impassioned imperative.

Since we ignore the physical rules, thermodynamics, gravity, etc. every time we move into the costly isolation booth of our suburban utopias, we are left with pennies and dimes to "buy into the spirit". . .The continent has the unflattering flavor of a gigantic pawn shop in which we try to buy time and respond to the distress signals of our consciences that are adrift in a distressed environmental sea.

Since the quick fix is no fix at all, it is going to take more time than we are willing to put up with. It takes time and effort to reform our habitat into the sound economy that coherence and sensitivity can construct. It will require time and effort on a scale that our corporate powerhouses can barely anticipate or simulate. But it must be an economy of the real at the expense of an economy of the practical; an economy of the desirable at the expense of an economy of the feasible-marketable.

Reality stares us in the face, and its mindless authority and power are unforgiving. We will not cope with it unless we find ways of transcending it while abiding by the rules of the game defined in 20 thousand million years of development.

Modestly and imperfectly I am trying to put into concrete and tangible form what I firmly believe is the call of the Urban Effect. It is the triggering of an urban-environmental re-form that is the comprehensive opening of the doors to environmental sanity, pollution and waste reduction, social and cultural enrichment, waning of guilt, effective coping with population inflation, proper and frugal use of our physical, biological and human resources, and last but not least, a tangible and exhilarating contraposition of *immense* nature and *intense* humanness.

Chapter Thirteen

Dichotomies

I make a distinction between envisioning things and being a visionary. To envision is not to be a visionary. The distinction is important because it points at the difference between realism and day-dreaming.

It might well be that for lay persons what is not hardware-oriented or discomfort-abating is what is the stuff of dreams, the non-practical. But among the stuff that dreams are made of is what is transpractical, the realistic, and what is extravagant, I would propose that the extravagant is the stuff of utopia, while the transpractical is hard-nosed realism and it is the creational event. Hard-nosed realism has a hard time to find a constituency because of its rationale that points harshly at what society wants, and about which society wants to deceive itself.

For instance:

To envision that the complexity of reality is an ever–self-incrementing process and, therefore, our action upon reality has to reflect and profit from such engrossment has not much to do with the visionary.

To envision that our habitats are or should be crucibles for complexification and as such must stand for a performance essentially complex and, consequently also "miniaturized", has again not much to do with the visionary. Miniaturization is the spatial parameter of complexity. If then there is a capacity to translate such need into planning-architecture, then what comes about is real-ization by way of which the conceptual (the complexity-miniaturization paradigm) transposes into implementation. The process is not visionary but inventive and hopefully creational. But, and this is paradoxical and crucial, it is also the first step into the nuts and bolts of existential coherence, congruence and fitness. Those happen to be the mandatory premises for equity and reverence.

It boils down to the hard fact that practical wo-man drowns into its own value-fiction, the realm of self-deception (visionary?) while realistic wo-man, the misnamed visionary, acknowledges the "mysterium tremendum" of reality and tries to be true to its own condition within it; the search demanded by the equity imperative.

I propose then that, contrary to popular classification, that I will be listed among the realists, equally apart from both the practical professionals and the classic visionary-utopianists. In fact, I would bunch together both practical and utopian types, as neither faces reality. Practical types, face against the brick, have no way to envision what a wall might truly be. Not only because a particular brick is too small a fragment but because it may not be the truest brick of all. The classic-visionary-utopianists avoid reality by dreaming of perfection and blessing, paradise on earth. They maintain and endorse a deceptive, false notion of nature (reality) and of her ways. One is mired in pseudo-analytical ways, and engages in perfecting an ever-voracious marketplace. The other afloat in the sweet and righteous light of clairvoyance and prophecy is disengaged in daydreaming and ineffectual drifts.

Then there is no real dichotomy between the two, inasmuch as both hinder the valid progression reality is entitled to, they foster waste, mediocrity and inequity at the expense of frugality, value and equity.

Taking "leave me out" as my motto à propos this issue of the visionary architect, I might still want to call the attention of the reader to the fact that reason is also the (unfortunate?) situation of knowing that the key to real-ization is hard to implement, because of the noise-pollution-confusion of the existing practical-utopian alliance.

Also to envision the possibility of some promising developments architects, planners and environmentalists should:

1. Guard equally against *the Adam and Eve syndrome* that proposes human history as the only reality worth

considering; and against *the Tao-Nirvana syndrome*, for which the unbounded and unquantifiable turns out to be limbo, unstructured and ultimately Philistine.

2. Guard against the false humility that suggests submission to the demands of the client. Interpretation entails endorsement. Endorsement without discrimination is acquiescence, and acquiescence is a bad omen for both client and for those who render them services.
3. Guard against the relentless onslaught of consumerism. Do we have to bury nature (and farmland) while we attempt to construct environments (habitats)? What about seeing the suburban diaspora for what it is turning out to be: the wasteland of a broken-down American dream?
4. Guard against the sweet song of peer adulation, if that means to bury the pangs of a conscience no longer at ease with itself.
5. Guard against technocratic incantations full of cunning but short on passion. They are spellbound by the quiet fury of "efficiency".

Science and history are pointing at the direct nexus between the richness of life and the complexity of its make-up. The rich life of a worm, compared to the "dreary fate" of a bacteria, is also pure tedium when viewed from the restless going about of the worm-eater, the bird. But to lead the life of a bird would be dementia for wo-man. The physiology and psychology of all those creatures reveals (as far as we can infer) the exponential growth in the complexity they exhibit in relation to one another, since the basic stuff (the atomic-subatomic particle-waves) is identical in all of them. What subtends the relative complexity that animates and sur-animates them (mind-spirit) is miniaturization. That is, the ever more subtle ways by which designing, programing, articulating, cooperating, metamorphosizing, etc. can "stuff" into the living envelopes occasions and events in numbers that quickly become unimaginable. We speak of trillions in enumerating the identifiable and distinct occasion-events within a human brain.

The collective organism, that which we come to see more and more as the logic and potentially explainable totality of earthly life (the Noosphere of Teilhard de Chardin), can in no way escape the same imperative, that of the complexity-miniaturization paradigm. What is promising in this "necessity" is that the "virtue" it might or should solicit from us is nothing less than a renaissance of the city; as that crucially necessary manifestation of civility, culture and spirituality essential to the existence of a healthy, meaningful society. Again I call the process that authors it the Urban Effect and I see it not as a limited and human proposition, but as the central drive of the evolutionary escalation from bacteria to trans-woman. This has plenty to do with the work of the planner and the architect. Our habitats are hopeless in historical and evolutionary terms (realism), if they do not author and reflect what is complex-miniaturized. As we have said the energy crisis is a powerful reminder for us of our spiritual-mental crisis as is the bird-worm-bacteria-inclined syndrome. That is why the suburban formula is not only a wrong but deadly formula. It is unfortunately the easy escape, the deception hatch, and we have been passing through it in greater and greater numbers during the post-Nazi war period. Any other consideration is a distant secondary priority to the imperative of complexity-miniaturization; even if our social engineers, our critics, our historians are ruefully unaware of it and go on pounding in the wrong nails. Such pounding turns out to be ineffective for both critics and criticized because it joins them indiscriminately into limbo while reality runs on undeterred and merciless.

Let me say a word about the complex versus the complicated. I find Webster's definition wanting: "Complex . . . something made up of or involving an often intricate combination of elements. Complicated . . . difficult to analyze, understand or explain."

The sharpest distinction one can make is the contraposition of an organism with a machine. To make the parallel as fair as one can, one must compare the simplest, most elementary organism, a bacteria let's say, with the most complex of all machines, the air carrier for instance. Discounting the intelligence put within its make-up, the carrier is a complicated event, the bacteria is a complex event; the carrier is technological, the bacteria is ontological; the carrier is a means, the bacteria is an end. Our intelligence suffices to invent, design and construct the carrier but it is not up to inventing, designing and constructing a single bacteria—not yet. One is processed, the other is living. One is outer-driven, the other is inner-motivated. One is simple-mindedly complicated, the other is complex. One breaks down, the other is self-perpetuating.

The human make-up is such that it generates complicated occasions in a reality that is complex. The complicated occasions are beneficial if it is understood that they are in a sense simple-minded concoctions designed to obviate weakness or want. It is hard to write by using a finger dipped into a pigment. Therefore, the pen was invented. But the scribbled page is a poor substitute for a living thought. The boat or the submarine are elementary systems compared to the whale. The complicated assemblage of the boat pales in coherence, design, efficiency, performance, when compared with the immense subtlety of design, the profound coherence, the incredible efficiency, the astounding performance of the whale. The complicatedness of the boat cannot be a paragon for the miraculous complexity of the whale. Our habitat is a compound of complexities (us, our social-cultural nexus and our pets) and complicated systems. The complicated system must add up to a synergistic proposition within which can flourish the complexity of the trans-personal and social-cultural milieu by developing the complex phenomenon of civilization. The esthetic is fundamental to the process. In fact, it might well be the sum of it.

Chapter Fourteen

Animism and Abortion

The pro-abortion/anti-abortion dilemma typifies the confusion vitiating our present model of reality. We can all agree that abortion is not just an unpleasant act but it is also a sorrowful one. Beyond that, there is confusion. At the outset I must say that I will not debate anti-abortionists who feel that they have a mandate to preserve life. I am in agreement with such a general statement. But I will debate them on their faith in a future that God has devised for life, be it embryonic or mature. And I will not support, nor would I have any ground to stand on abortionists who believe in God. They are a living contradiction, well meaning but fuzzy of mind. God-fearing people are right in labelling abortion manslaughter. What is to be asked of them is not a recantation about their perception of abortion but instead a reconsideration of the reality of their God, of any God in fact.

If God exists, then life generates from God. Therefore, life, all life, is sacred (and pre-destined), bugs and cobras included, even the fetuses and monstrosities that sometimes accompanying them included, billions upon billions of male spermatozoa included. The animistic (God creator) logic is iron-clad logic. What is shaky is animism itself, not a model that one constructs with it, the abortion question included.

For pro-abortionists who believe in an animistic reality lose the battle in the wargames room. "Abortionist" cannot have it both ways. The way of animism, is that of a reality willed by a God or by gods. The way of the abortionist proclaims the wo-man's right of deciding as to what errors or distractions of God should abort. It is an incoherent position. The most one can ask for is some sort of God turn away while I abort type of justice.

If *all* manifestations of wo-man's life are God's manifestations, that is, they are sacred. An impregnated egg is a manifestation of life. Therefore it is sacred even when it is defective. This is simple unimpeacheable logic. It doesn't date. God has spoken. We must bow down humbly and prepare for our new life.

The "abortionist" has a hard choice to make. "Go with God" and stop interfering with life as God wills it, or be responsible for what life can be and put God, a wishful notion, aside. If God is, abortion must go. If God is not, abortion is a sorrowful but at times legitimate act that life might want to be responsible for.

All the debates, the accusations, the invectives are off target. *God and ghost* are the center figures, not child, mother, father, society, etc. If God is, abortion is a sin against life; if God is a ghost, then abortion is here to stay, along with other sorrowful realities.

The animist (God's believer) has no use for the distinction between *potential* and *actual* because in God's mind the two are indistinguishable. That which is potential in the temporal world of wo-man is complete and unfolded in the timeless reality of the Godhead. Therefore, the potential is in the actuality of the developing organism tightly packaged in the blob of flesh growing in the womb. The embryo is just unfolding its own *already-present* reality. The believer sees in the tiny blob of flesh growing via parthenogenesis not a potential person, not as an aggregate of cells, but as a person. It is the initial phase of a process of creation that is capable of lasting for the next 60-90 years, to a point where *the* person is fully exorcized from matter. For a believer, the distinction between an apple seed and an apple tree is a mystery perhaps, but it is so only from a human perspective, not from an omniscient or divine one. Both squashing the seed and felling the tree is sinful. They coincide since the one *is already* the other.

The problem of the believer is with the deceptive nature of time and with the transformation-creation that time causes. Once reality is divine, there is no gradual injection of sacredness within its manifestations (organisms) and their development. The speck of flesh is statically, mandatorily, despotically, inalterably sacred; mindless and impervious to the fact that sacredness proceeds from the sufferance of time passing, physiological growth, the appearance and development of intellection, etc. Thus the cutting off of an embryo from a placenta wall is no different from the execution of a criminal. Again the animistic rationale is. . . rational.

What is open to question is not a rationale of effect but a rationale of premise. If reality belongs to God, its creator, then the effect is that sacredness is intrinsic to basic organisms. Development (creation) is therefore a revelation of an existing reality. Beethoven's embryo was not a remote possibility. It was Beethoven. If the existence of God is not up for debate, then the anti-abortion position is *unshakeable.* The Catholic Church is right and abortionists all cast adrift on flimsy rafts of convenience, in the face of the monumental weight of God's unflagging will.

But drop God from the drama and what do we have.

1. *Self-responsibility* surfaces and demands to exert itself.

2. *Creational development (evolution) dislodges revelational unfolding.*

   What is the worth of something that is intrinsic and actual. An embryo is the equivalent to a few multiplying cells following a genetic dictate. It is not contingent on something (an embryo) it might have become if it had not been aborted. If the embryo of myself had been aborted, I would not have been removed from reality, for the basic reason that the embryo was not me. It was a rudimentary pedestal upon which I could have been generated by time and circumstances. With the embryo aborted, the hypothetical me would have remained exactly what it was at that point: One possible avenue for life to become something. This is as true for spermatozoas destroyed in the womb as it is more basically for the many ova that have been denied access to the womb and thus to the development process.

   This is exactly what distinguishes evolution from revelation. The substance of an evolutionary-creational process is always where the process is any given moment. The meaning of a revelational reality is not found in the cumulative substance authored by the process; but in what is revealed, what is unchangeable and what is out of our hands. . . In this case an embryo is only a person in process of "unfoldment", or self-revelation. The folds are all there but hidden because they are stacked up.

The absurdity here is that even though an embryo might be destroyed, the not-unfolded reality must still exist somewhere, in a limbo, in a kind of suspended reality, perhaps even in the mind of God, and in a nightmare of a God-fearing abortionist. This robs life and reality of its presumed sacredness. The abortion issue becomes mute within the stillness of a reality fatally determined by God.

3. *The sacred worth of an organism is proportionate to its development* because such development is what makes the reality of it. The genetic matrix is the fruit of billions of years of evolution. It is ineffable on its own terms. But regard the unbelievable (and wanton?) largess by which genetic boxes are thrown around at all levels of life, on both male and female sides, both before and after impregnation.

4. *The natural anguish of a woman who submits to abortion must be measured* against a background of the increasing worth of the organism growing within her, but outside of the sinful malediction of a foreign and deterministic God. Such anguish should suffice to remedy the misdeeds and sorrows of abortion.

To advocate legal abortion is a lost cause within an animistic (God believing) model. In an animistic scenario the intensity, (the sacredness of life) is in everything—everywhere; and that time that terminates life's processes is *always* evil. For a barely conceived embryo is not a potential consciousness. It is an individual with complete rights. Since its development is an unfolding rather than a continuous self-creation, all of its sacredness is present at the onset.

One could propose a law whereby spermatazoas and ova since they have attained the 50% level of personhood should then have a full and unconditional right to life; and that the remainder of our living kingdom should fall within this same right since worms and bugs, algae and fish also participate in that unfolding sacredness pointing to spirit. I repeat it is not by an extrapolation that this model becomes absurd for the model already is.

The creational (evolutionary) model is far different. It proposes that any beginning is in a sense, a true beginning with a single asset—a genetic matrix. A genetric matrix is not personhood. It is a grid capable of painstakingly imploding around itself its own personhood. This happens by degrees starting from the moment of conception. To dispose of sperms and ova is an automatic, logical decision deprived of ethical overtones (at least at the logical level of this discourse). To dispose of an embryo one minute old is a decision almost totally devoid of ethical overtones. For days and weeks the creature is a speck of tissue slowly becoming an organization of perceptions-actions. Before it achieves anything comparable to the sacredness of an adult mouse, it will take many weeks. Even then look what we do to adult mice.

By mistaking potentiality with contextuality, we vitiate the occasion. Consequently, the few days-weeks old embryo "Voilà" is a person. In trying to animate everything with the spirit of God, the processes of growing, suffering, developing, meaning making, and valuing drown in a sea of righteous, embryological and even proto-embryological events that attribute personhood that is truly abortive and incapable of becoming (self-creating instead of unfolding).

Any embryo they say, might be a Beethoven or a Mother Teresa. The fact is that between their embryos and a grown reality of a Beethoven or a Mother Teresa there exists exactly what it took to make them what they are. Aborted embryos are not aborted Beethovens or Mother Teresas or person at all for that matter. They never were persons; therefore, they could never be aborted. (An aborted flight is a non-flight.) What is aborted are granulae of flesh that contain the potentiality for becoming specific persons. But the persons do not exist before they becomes such. We may cry because of an aborted anticipation, but we do not cry for that which never was. That would be absurd. A genetic matrix hardly makes for humanness, per se. It is a templet to be used by a developing personification. At the beginning there is the "wiggling pseudo-fish" (ontogenesis as a portrayal of philogenesis).

If we travel the road of animism (God), predestination points the way, and the future is present in the past; that is, the embryos are murder victims since in the eye of God (the abortionist of time) they are persons. It is up to God to justify all the toil and suffering inherent in making something that which it already is. Animists see this in any seed disposed of: a tree or in any animal aborted. *They must then appreciate the fact that reality is an immense sea of aborted lives punctuated here and there by those rare organisms that have escaped the norm (abortion).* As for the preconception phase, where did God get the crazy notion that in order to create one person, he had to get into a messy mass-murder, supermarathon type ending with the total extinction of millions of spermatozoa less one! This God must be kidding or God-fearing people must do some homework.

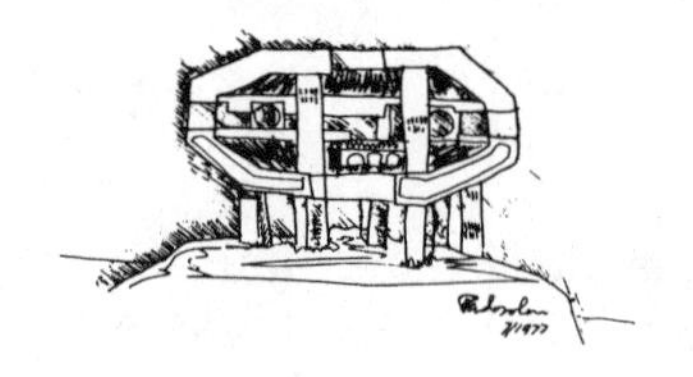

Chapter Fifteen

Sky Cities

Four years ago I wrote, "If science is correct in determining the earth to be composed of the same stuff which makes up the cosmos, and if life is the technology through which such stuff becomes animated, then life, which now appears to be the exception to the rule within the physical cosmos, could eventually become the rule. If, in addition to being feasible, this potential animation of the cosmos is also desirable, then the responsibility of life lies in the transfiguration of an immensely powerful physical phenomenon into an immensely loving, spiritual one. An eschatological imperative.

"There will be renewed religious unrest, caused by the space probe, focusing on eschatological issues which will embrace social, environmental, cultural, ethical, and esthetic concerns. All operate directly upon the human condition in conjunction with the questions of health and genetic preservation.

"But the eschatological concern will be largely unspoken and intentionally brushed under the rug of hard facts and technopolitical imperatives. And yet, the stakes are frightfully high; we must face what we are about to plan and implement. Under the pressure of scientific and technological "progress" stimulated by the space venture, the eschatological concern will give rise to new or pseudo-new theological models. Thus, as I perceive it, the probe of life into space is, ultimately, not a technological, political, or economic problem but a theological one.

"The eschatological implications of space colonization are fundamental and critical. I will consider the following aspects:

1. *The eschatological concern*: the question of the ultimate aim and purpose of life, the eschatological paradigm as such.
2. *The genetic concern*: the splitting of the human species into subspecies (which has prehistoric precedent) as a direct consequence of space being invaded by humanity.
3. *The urban concern*: the space probe is the urban probe on new grounds; therefore, the urban question looms ever larger in the destiny of the species.

"Depending on how these questions are approached, hope or despair will emerge.

"In despair:

1A. We see ourselves as well-worn sorcerer's apprentices, incapable of halting the plunge into technological hubris that more and more forcefully brings upon us the wrath of our indignant Father, the Lord, and our merciless expulsion from nature's bosom.

2A. The human species—abandoned by the Lord, Providence, and instinctual wisdom—will tear itself apart under the stress of new (evil) environs and spawn inimical subspecies which are foreign to each other and will find their own nemesis in specialization, genetic or otherwise.

3A. The human species will make an evermore compromising step into the urban syndrome, seen as the sum of all evil, since space colonization will be directly informed by those conditions which per se define the current urban context.

In hope:

1B. We are making (reaffirming?) a promethean commitment to the spirit by taking action to release it from the gravitational vise of the earth and by opening the cosmos to urbanization that is, to logos.

2B. The human family recognizes its genetic and other limitations and willfully seeks morphologically new cognitive forms for the purpose of outfitting itself for the immense journey into the spirit via the flesh (mass-energy), the process being a lengthy series of transcendences.

3B. In stepping off the earthly landscape, man is turning (by necessity which will become virtue) toward a frugality of environs and hardware specific to urban condition, toward evermore crucial transphysical processes. In sum, man will opt for the self-containment of his habitats, their inward orientation, the cooperative and interdependent nature of the social and cultural texture, high density performance, the imperative of integrity and self-reliance and, finally, the complexity and miniaturization of the milieu. The space city will, therefore, be an unequivocal test of our readiness for the vertigo of a momentous step toward spirit.

"Then on what can be defined as the threshold to the infinity of space, the human species is going to make critical decisions. Some of them will be unconscious and irreversible, some will be conscious and crucial. First, if the space colony venture is to be developed, we must seek consensus. In an undertaking of this magnitude, it has to be a transnational consensus and it must be knowledgeable. This will be impossible if the pioneers and promoters themselves are less than clear about the scope and impact of the enterprise." (From *An Eschatological Hypothesis*, Doubleday)

All Cities Are Sky Cities

This is not meant as a paradox, but to emphasize the common nature of all habitats. They are local order. As such they generate disorder elsewhere (entropy drift).

The order they foster partakes of the dynamic, gravitational, thermodynamic, relativistic time-space reality of which they are part. They belong to the same sky, the prevalent mode of our mysterious universe. They are life pods within a non-life scenario. They are each an innerness painstakingly asserting itself within an immense sea of outerness. They succeed in doing so by feeding on such outerness. Such feeding might someday achieve such proportions as to consume all outerness by the—then all-pervasive—innerness. Some religons see the spirit as metamorphosis of "matter", all matter. They propose this as a permanent and everlasting condition in their contention that tangible reality is purely an expedient end (insubstantial), a deception. Such is the animistic revelational model.

But it might well be that the living phenomenon is order wedging itself into the relatively inertial spin of physical things. The soft underbelly of revelational reality is that it wants, it needs, to see order in the preordained, the providential nature of animism. Therefore, there is no such thing as a wedging of order into reality but only an uncovering (revelation) of order to the human context. The preordained nature of animistic reality dismisses process. The demise of process in turn freezes the creational instability and its unforeseeability. For animism reality is, has been, and always will be the self-same and time is only the evidencing of it just as one evidences things filling an attic by moving the beacon of a flashlight eratically here and there. But the attic is ultimately a cemetery, the least ordered of all things since an object there is an empty shell, drained of will, disorderly—in fact disorder as such. I would discard the animistic-revelational mode because for it all things, big and small, are fatally predefined, sky cities included. If for small things one can brush aside the eschatological dimension, then for sky cities the eschatological dimension runs through their axis. Without an end in sight, an irreducible goal, sky city is just another gadget, a technocratic caprice and a costly one. The sky city proposition, as the proposition of all habitats, stands or falls on its capacity for spirit-making (and entropy generation). There, at least in evolutionary terms, is the sky city dilemma.

How (per se) disorderly in the entropic sense can habitat become? Any ghost town can tell. But there is an even deadlier kind of disorder, the technocratic disorder. It is so because technocracy is frantically spinning and spinning while piling up physical disorder in its vicinity and spiritual chaos in itself. What sky city must dread most is such a condition: the physical disorder produced by high technology (second law of thermodynamics) and the naught easily nesting in its convoluted structure.

Eventually, sky cities will epitomize frugality by the minimal ecological mass they will need for support. But they will be an organism (as association of creatures indissolubly bonded) drifting in an ecological vacuum. That is, a brand new organism importing, from galactic distances, mass-energy to be made into life. Accustomed as organisms are to the immense variety of this planet and to the endless articulations such variety offers, the sky city will be a "deprivation city", a poor cousin of the frugal city. Perhaps the answer—a dangerous one—is to put body-less minds in the sky city to match the ecology-less environs of the sky city. Such a thing would be an exalted, crewless, space probe, until the day when brains (physiological organs) can be isolated from the body (as we have it now). This will mean a de-escalation of physiological complexity to match the de-escalation of the ecological complexity of the sky city. If we knew a little more about effemeralization, we could perhaps be less grim in forecasting. Since I put my chips on effemeralization, I endorse sky city but only if the paradigm of complexity-miniaturizations-duration is observed and practiced. For now we call for pseudo-ephemeralization, that kind of efficiency reproduced by a simplification, a direct consequence of the analytical mind having the upper hand on the synthetizing mind. Technology as we practice it now is the child of simplification. The grandchild is pseudo-ephemeralization, the potential nemesis of sky city.

As one can see, I am not even addressing feasibility because to my perception our conduct should not be dictated by feasibility but must be driven by desirability. The meaning of sky city must be found before we plunge into its making. The scourge of technology is the magic hold it has on an imagination which has not yet been bridled by meaning. Ever since the human cortex has begun to anticipate, abstract and plan, the separation between feasibility and desirability has been the number one problem. Suffice to say that conflict is most often present when and where feasibility has not supported desirability (as the "good of mankind"), but has encouraged the compulsions, the ego of this or that person or group. The chasm between desirability and feasibility is where consumerism casts its dark shadow: its inequitability. If sky city as prototype can stay out of such shadows by being in the realm of protodesirability, sky city as a marketplace entity will have to cope with its own propensity for "explosive" feasibility and the hubris or mindlessness it entails.

I am closing by quoting myself again: "Urban experimentation should be the top priority for any concerned society. It is in urban well-being and liveliness that one finds the frontiers of a fair, equitable, civilized society, a society with a future. The rejection of experimentation, aside from being antidemocratic, and *a priori* a denial of possible freedoms, is also arrogant and intolerant, growing out of the self-adjusted, 'reasonable', and not modest stance of mediocrity, the other 'emergency' of democracy.

"We must recognize that there is no integrated effort really concerned with the future; that is, with us and our offspring.

"We must recognize the imperative to pursue that effort in order to keep open options for the future of life, ours included.

"We must recognize that the unfocused nature of the technological onslaught is steamrolling friends and foes.

"We must recognize the political impossibility of asking voters to endorse an effort that will bring massive dividends only after most of them are dead.

"It then appears that, in spite of the existence of many strategies, still another must be developed: the declared, intentional, determined pursuit of exotic ideas that offer the opportunity of testing, verifying, and assessing notions, systems, and situations *now*, that may well become run-of-the-mill in the next century, twenty years or so away.

"The exotic seed is seen as a tri-pod (pod as a protective envelope, a clustering of life) organism: Pod 1 is an urban system for land; Pod 2 is an urban system in the seas; Pod 3 is an urban system in space. In developing the technology and know-how for space it becomes more and more evident that the simulation of space-colonization falls short in the most vital area: humanness. Here simulation is ineffective. It is, therefore, imperative to actually experience, as nearly as possible, environmental conditions that may be indicative of the critical context of space. The land pod could offer such environmental conditions. The sea pod is located somewhere between the land pod and the space pod—closer to the former, inasmuch as the aquatic environment is still terrestrial. All of the pods, however, have one most significant element in common: they are urban organisms. As such, they belong also to the backbone of evolution." (From *Fragments*, Harper & Row)

Chapter Sixteen

Teilhard and Metamorphosis

If a drop of ink falls into a glass of water it diffuses evenly and grays the water. In a large enough glass of water the graying will become imperceptible. Dipping a pen in such a medium in the hope of writing a message, any message, is futile. Nor can the message gather itself within the body of water. The word (verb) might well be there, but for all practical purposes it is nowhere. Particles of unadulterated ink are yet to be found, but the segregation between them is such as to frustrate any performance. The medium, water, has killed the message, ink.

The problem with animism for which spirit—ink—saturates matter—water—might be seen here. Something might well be in matter which speaks spirit but the voice is so feeble as to be lost in the bulk of matter itself. Only the immensity of it makes its mystery excruciatingly "vocal" when and if there are ears to hear. The living (phenomenon) seems to be doing the impossible; it recollects and organizes the isolated particles (of ink) into images (organisms, the verb). In the process it pays a dole, the entropy tax. For each creature coming to being, there is a dissipation of energy that seems to augment the enertial bulk of reality. More groupings of ink particles go for greater weakening of ink presence in the water. Disorder (weak concentration of ink) surrounds order (concentration of ink). The faster the transformation the faster the coming about of entropic death!

The "water" is reality hosting the *l'infant terrible* (ink) that in paradoxical defiance of feasibility and determinism, keeps creating vortex or consciousness within its maternal body.

The point I'm trying to make is that the more life is spread out and diffused the more it becomes mute and to all purposes impotent. In fact it dies off. There is no ink in Lake Michigan even if you pour in a supertanker full of it.

The ink, drop or bottle, is the urban effect and its epitomy in a remote future is divinity, the verb. Not just a piece of information but in-formation. Not a fragment of grace but gracefulness. Not a work of art but the esthetic. Not a technological exploit but absolute articulation. No message dispersed within the media-matter but the message pure and all-inclusive. (The whole body of liquid has become ink.)

For me Teilhard de Chardin's crucial tenet of convergence stands as the imperative of intensity: the density of ink to produce the message of consciousness to a degree that sees all limits of consciousness done away with, resolved. A divine consciousness.

The urban effect *personifies* the Teilhardian paradigm of convergence; a paradigm I trisect into a component of *complexity*, a component of *miniaturization*, a component of *duration*.

To write a living message, single particles (ink) must collect themselves together in sufficient numbers and sufficient proximity to become visible, and interactive; that is *miniaturization*. But the message is not a single utterance or shout but instead the polyphonic, orchestrated music of community; *that is complexity*. This message, this music, this life, make time irreversible and potentially unhinged from irreversibility itself. *That is duration*. If, then, time becomes non-reversible, as classical science proposes, or irreversible as the new physics suggests but becomes also retrievable as equity demands, then somewhere in a future too distant for conjecture a divine body-point will contain in itself the full experience of cosmo-genesis and evolution. As such it will be the ultimate seed of such experience.

The imperative of equity is that equity be retroactive. Without retroactivity, from any moment in history (evolution) all that comes before is, at best, a means to a hypothetical present equity. At worst, it is a meaningless sequence of random events without a head or tail. Only a reality that has created a condition in which all the inequity has been conquered is an equitable reality; it is easy to see that such reality has disposed of time, space and mass-energy and has to be the receptacle-tabernacle of a message which has no residuals of media.

The polis is the gathered occasion where individual drops of consciousness come to interact polyphonically, thanks to the institutions invented by the mind in a long evolutionary stretch. The management of the polis, its policies, its political construct is not the purpose of the coming together, but it is a necessary part in the process of developing and running it. Such a construct can be read occasionally in the best circumstances in the layout of the bulk, the masonry of the polis itself. It optimally should be so if it is true that we are the children of the environment; that is, of "Mother Nature" and of the constructs her children do on and with her for the purpose for fostering their development.

The polis-urbis is not expedient but ontological. It is the extension of the individual single urban effect constructed by a single *organism* into the collective urban effect of the town's organism. Similar in methodology *but at a different logical, mental, emotional, psychological level* than the ant hill or the termites' nest. Not just thousands of brains going about the communal tasks of survival and reproduction, but also thousands of "free" minds going about a communal task of transcendence. Perhaps there is no conscious acknowledgement among people that analogically the urbis is organism. Its existence and reality is the *sine qua non* for the survival and development of peoples. (One difference is that the greater versality of wo-men makes for interchangeability, in migration and substitution among the human aggregates. But this is in general a modern development. Peoples used to be born, live and die in the same community.)

The nexus between our project guided by the urban effect, and Teilhard de Chardin is not casual. Teilhard spoke eloquently on the concept of complexity and the notion of convergence (Omega) and it is with such a referential grid that Arcosanti and my work is seeking existential and contextual answers. Such a grid cannot escape the demand for an eschatological focus.

I will quote myself and reiterate the importance of the Complexity-Miniaturization-Duration Paradigm.

Complexity-Miniaturization-Duration Revisited

If love is as I think it is—the necessity of seeing meaning in all things big and small—love's enemy is noise. Noise is the innate incapacity of space to "let information alone" and the innate defect of time to allow information to age. The consequent distortion and obsolescence imposed upon information makes it mis-information. If, as a living cell in a composite organism, you misinform your biological neighbors, other cells, you and the organism develop pathologies. The same goes for neighboring organisms informing each other. We suffer a pathology of distorted information. Luckily, the foundations of our consciousness are in better shape. Physiologically we do all right. Our physiological information mechanisms are subtle and flexible thanks to a miraculous ephemeralization of time-space noises. We are packaged tightly and as a consequence we are physiologically well-informed. Being well-informed we are in-formed and so we per-form. We do in function of our form. (We love the form which is ourselves.) And we are a performing affirmation and a perpetual confirmation of the complexity-miniaturization-duration paradigm. The immensely frugal utilization and the living organization of mass-energy, time-space goes for the effusion (dissipative act) of consciousness. By way of complexity-miniaturization we move out of time and become tough; that is, a durational event, tiny bits of everlastingness predisposing themselves in the mosaic we now call the past for perhaps that which could be the miracle of miracles, all past becoming an endless, everlasting present. Mega-miracles aside, the future of the foodchain evolutionary process goes on by the rules of the paradigm or is doomed. Complexity-miniaturization-duration fits:

A. The mega scale of things where the diffusion resulting from the Big Bang (the Big Bang has the radiance of an abortive attempt by reality to conclude itself into its own semen. The semen, not being fully conceived, aborted in a way that only something as immense as the cosmos can abort. Self-annihilation by way of vaporization. Perhaps what float around and excite the animists are shreds of spirit miraculously surviving the horror of the Holocaust—Big Bang) becomes the scenario for constellations of events, the implosion of scattered particles in loosely organized systems, galaxies and their components. That is, from that which is explosively enlarging, minutiae organize themselves in proto-complex, proto-miniaturized, proto-duration organisms, galaxies and their components.

B. The micro-scale of reality where our intervention as investigator makes giants out of our instrumentation intruding into the ephemeralized dimensions of the complex sub-atomic universe. In it we are in the process of discovering how much happens in infinitesimal spaces for infinitesimal time and the ability of such mini-events to "create" enormous outpourings of energy, the stuff of proto-consciousness.

C. The macro-scale of reality where the endless inventiveness of life makes both cosmic and micro-cosmic events bow to it and become means to the transfiguration of their reality, into consciousness and mentation: Matter becoming spirit on the bridge that matter metamorphoses itself in, all along the process.

D. Whatever form of divinity one sympathizes with to the exclusion of those forms endorsing maximal entropy, that is, the de-formation of chaos and the thermodynamic death of reality that failed self-transcendence. It bonds matter to mind in seeing one fathering or mothering the other, moving on a path which can be fatal but does not necessarily have to be so.

Is there a quantum jump in store? Perhaps yes. The technology (bio-technology) of the living might have reached its own limit, the macro-universe of living tissues. The technology of the "non-living" micro-physics and micro-chemistry are in full evolution. Each day brings in a new implosion of circuitry and memorizing. Work is being done in combining biological miniaturization with extra-biological miniaturization. Once technology reaches parity with bio-technology (living stuff) the quantum jump might be within reach.

It could be an event as magical as the appearance of life; an immense leap in the womb of matter to inseminate it and not just with the dynamism of body but (again and) this time with the radiance of mind. Then the cooperation of macro-biology-conscience and micro-physics-logic would deliver trans-micro-consciousness-intellection, the next evolutionary step in the complexity-miniaturization-duration universe of the Urban Effect. The Urban Effect is that which is brought about by the self-guiding and operational paradigm.

If one were to speak of a Teilhardian tradition, one could say that the effort of Arcosanti is in that tradition. For what is happening at the site in central Arizona I will refer to a statement written in 1979 on the Arcosanti Project, an urban laboratory.

What Is the Arcosanti Project?

It is a search and a research within the Complexity-Miniaturization-Duration paradigm. Search covers two areas. The first, appropriately, is a quest for that which is hidden, for that which is unproven or unexplained. This is also the definition of research, although the emphasis is somewhat shifted. In the second area, "search" is a misnomer. It is the quest for that which is not. In this second ground, since the future is not, search means the construction-creation of the future.

Arcosanti is search and research in the field of environment and habitat (the field of civilization).

Oceanography, plasma physics, cancer, big bang, metal fatigue, ozone layer, sociology, taxonomy, statistics, etc. are all fields of research (seeking for answers), so too is it for Arcosanti, but Arcosanti also searching for that which is not.

The urban phenomenon, engaging about two-thirds of mankind, is a mental-ecological compendium of many, many-faceted processes. Arcosanti seeks resolutions realizing that the complexity of the subject, far from being a deterrent for the search for answers, speaks instead of the fundamental nature and value of this search. The project is by the modesty of its assets, a simplified, ambiguous, tentative, assessment of the urban phenomenon and is a process attending to it. It is an urban laboratory for search and research.

The simplified, the ambiguous, the tentative are exactly the ingredients of research.

> *Simplified*: the isolation of specific occasions (heat transfer, for instance).
>
> *Ambiguous*: the indeterminancy of its niche within the larger system (cancer, for instance).
>
> *Tentative*: a temporary, hypothetical position and nature of the niche (is cancer viral, genetic?).

There would be no research, no need for it, if *wholeness*, *clarity* and *exactness* were the ingredients of a phenomenon. Those can be only, and only rarely, the end-products of research. *We have been and we are very clear about that.*

But there is far more to it than that. Simplicity (limitation), ambiguity (dialectic), and tentativeness (probing) happen to be also the fundamental characteristics of reality. They squarely legitimize the project, *if and inasmuch as the project has some clear notion of where the emphasis of its search should be*: Energy, resources, food, pollution, segregation, social-cultural intensity, self-reliance, autonomy, identification, pedagogy, etc. It should also be clear that the *economic* speculation per se—a post-laboratory ingredient—is secondary to the *vital* speculation—its validity as search and research. Willy-nilly we are dealing with what the mind of wo-man has called the *mysterium tremendum* (the tremendous mystery) of reality and life within it. Once more, *we kid ourselves if we think otherwise* (seeing only our personal troubles and the world-wide troubles and tragedies).

It is quite understandable that individually we might like to work with a clean, clear proposition, of our personal uncertainties, vagaries and irrationalities. This need is filled by faith, parochialism, etc. But the fact remains that to purge reality of the ambiguous and the tentative is also to purge ourselves out of it. To say that it would be much preferable to have a neat, ordained task with all the chips securely identifiable, etc., is to reject the very reason and purpose of research, especially if it is urban research which is the epitome of complexity, ambiguity, and uncertainty. This is the mistake of the determinist who, discouraged with the dumb-founding complexity of the frog, tries to learn and explain its physiology by playing with a clock-like or robot-like facsimile.

To deal with certitudes is to exclude the need for search-research. Las Vegas of all places is a good, if paradoxical, example. It is statistically unavoidable that the money left by the gambler will be, in toto, many times greater than the money he takes away. There is no gambling in Vegas' profit story as long as the suckers keep coming. The gambling of Arcosanti is a different story. *It does not come out of the certainty of escapism. It wants to deal with the ambiguities of reality.* The Urban Effect is the summation of all of the ambiguities of reality.

To go about it with some clear and simplified notion concerning its structural, logistical, energetic, managerial, environmental, social, and cultural facets is not a contradition. It is also not, even remotely, to deliver the golden egg. It is an attempt to clarify the intrinsic and "beautiful" ambiguity and complexity of living and to endorse its unavoidable transformatory power.

The longing for that which might be truer, for the reason why it does this or that drives us to research, as well as search. *It is at bottom the least quantifiable of all tasks inasmuch as it is the task which seeks quantification. It is the least foolproof of all tasks, inasmuch as it seeks the rationale of things,* of those things of which it is one. As if that were not enough, it also dwells on creating that which is not, an "exponential uncertainty". If it were to be, or pretend to be, that which it seeks, it would then be senseless, insane, moronic. *It seeks the rationale of something because it is on the outside of it. It seeks the quantification of something because it has not got the numbers of it.* It is a quest for what it does not have. It is in addition a time-change-creational process, a mix of knowledge, sensitization, anticipation, unpredictability.

Chapter Seventeen

Teilhard and the Esthetic

Is there a Teilhardian esthetic? I believe that there is, provided that an esthetic is a real aim of reality in the process of creating itself. A "beautiful divinity is unheard of. Can such semantics create our destiny, our escatological consummation?

As a category in the conscious world, the esthetic has been annoyingly unruly from time immemorial. It does not fit most of the respectable models of reality. Looking at this question from within a discipline, such as science, philosophy, religion helps very little. The esthetic manages to survive as an "enfant terrible". . .Yet in my perception it is the infant destined to embrace all and metamorphosize all—and I mean *all*—of reality, the cosmos.

At times, one goes on the assumption that beauty is an attribute of divinity and that consequently the religious person is a seeker of beauty. Facts do not seem to bear out such a rationale. Often a religious person speaks of the beauty of the soul or the beauty of grace. And often it is hard to separate esthetic pursuits from other rather despicable actions. We remember national socialism's infatuation with Richard Wagner. Is piety, love, compassion and the divinity itself beautiful? One then can ask whether God is loving, compassionate, graceful, because God is beautiful? Or is beauty itself defined as love, compassion, grace? Or is beauty some sort of ambiguity, a now you see it, and now you don't, with a shroud? Or is it a modulation that floats among objects, beings, actions? Is beauty a real category or is it an additive for stirring up morals, prayers, political actions, social commitments, compassionate responses, love, etc.?

My contention is that God and all God's attributes are but aesthetic manifestations. That is, the aim is not God but beauty. That is to say that reality gropes toward the beautiful and divinity is an attribute of this. Or to give a shocking example, it is not that artists are an instruments for the glorification of church and God, rather it is that religious institutions and their simulations called God, are at the service of a beauty that the artist tries to create. This is the evolutionary/cosmic process that I call esthetogenesis. Esthetogenesis comprehends but is not comprehended by compassion, grace, equity, knowledge, perfection. All those are instrumental to its genesis. The esthetic is the compassionate beauty, the beauty textured within reality when and after it has been filtered and recreated by the anguish of mind.

This is not a new proposition but it is still a radical or even heretical one that needs clarification to say the least.

I hope that I can give this proposition some coherence, make some point in favor of it on the next few pages. I will do so by again bringing Omega into the picture. My definition of Omega is somewhat different than Teilhard's. I see Omega as a hypothesis that proposes the only possible achievement of perfection. The achievement is a totality that is all-comprehensive in time and space and as such is equitable, and by means of its creational genesis, esthetic. As genesis and as totality Omega is the seed of the cosmos. It is the seed that generates from the endless metamorphosis of evolution. It is a seed that is the genetic matrix of its creator (evolution) and the trigger for resurrection. Since a seed is in a way the resurrection of the parents, the ultimate seed, at times consumed, is the resurrection of all of reality.

Wo-man's mind has struggled for thousands of years trying to rationalize the esthetic. All definitions are untrue inasmuch as trueness demands completeness, and completeness is not of this world (but of the world that is to come, of which this world will be but a part). But if they are untrue, they are not useless or mindless. They reflect the environment that they come from; or they reflect the response or reaction to the environment they are part of.

I propose the divine to be the prerogative of the infinitely beautiful. In fact, at the end of the tunnel of time it will be impossible to distinguish the beautiful from the graceful or from the equitable, lovable, or harmonious. Omega will be all of them (or it will be a deception) and all of them form the concluding act of estheto-genesis, the creation of the Omega seed of the cosmos (reality becoming); that is, Omega *will be* all of them or will not be.

There is then a provisional kind of beauty, as there is a provisional kind of grace, and a provisional kind of equity, etc. Inasmuch as they are perfectable *they are not perfect*. Then, for instance an imperfect equity contains a denial of itself. Inequity is part of the perfectable equity. The only equitable equity is the equity not available yet to reality, therefore reality is inequitable. Would this mean that Lake Michigan is black because it contains a drop of India ink? Yes, in the sense that there is blackness in the lake, even if it amounts to only one drop of black ink. Perfection is exactly what the blackened lake is not, in the sense of water unadulterated by ink. It is also quite clear that the inequity of reality amounts to tankers full of black ink, in a lake of a reality that is made up of pasts creating the future by means of the endless stream of present upon present.

In this endless process the seeking of equity and justice comes about with the appearance of a wo-man's mind that proves more engrossing than their earlier versions. The human mind (-body) cannot find satisfaction merely in fitness. It must seek after equity and compassion. Nor can it find satisfaction in being just. It must seek justice and love. Reality is impervious to such "transcendence". Therefore, A) the mind invents that which embodies equity, compassion, justice, love and calls it God. or B) Wo-man "depicts divinity by means of the esthetic, not only by way of a subject matter (such as the crucifixion for instance) but by stance. Therefore, my definition of esthetic:

Given a reality that is intrisically unknowable and whose development, evolution, is intrinsically inequitable, the human soul is prey to an inextinguishable anguish that is impervious to the blandishments of science, medicine, psychology, religion, philosophy, politics, etc.

What can to a certain degree transcend such a reality and transform the inequitable is the esthetic act. Since the unknowable and the equitable are transrational and transtemporal, it is the act that metamorphosizes them. Such esthetic action is more than an anticipation of perfection, grace, divinity, it is a splinter, a fragment of them. It is a piece of divinity; a particle of the full and completed esthetogenesis.

In fact divinity is the end result of a (total) unification of all such esthetic dust into an esthetic genesis/denouement. Omega is *the* beautiful.

While art transforms matter, is matter transfigured? Religion seeks the transformation of matter by way of myth and prophecy.

Religion designs all sorts of simulations of Omega. And they fall constantly short of the target. It could not be otherwise. We cannot even describe, or simulate an existing dust particle of the esthetic because the esthetic "thing" is indescribable. How could we describe the summation of all beauty, the non-existent beauty of the non-existent Omega?

If the beautiful has its apogee in the full esthetogenetic process, metamorphosizing not sections of the cosmos but the cosmos *in toto*, the esthetic is fundamentally "manipulation" (metamorphosis).

Then what distinguishes technology, manipulation via science, and esthetic "manipulation" via anguish? The first is instrumental for the creation of the second which is the aim: The metamorphosis of "matter" into beauty/spirit.

Here is where one can propose a Teilhardian illumination of sorts. As "matter" is not burden but media, the transformation-metamorphosis of matter is not expedient but substantial to the process of "divinization". Then if science is a dialogue between physicality and mind (Ilya Prigogine), then something more must be going on besides dialogue if "spirit-God" is to eventually create itself. This something more is "manipulation"; that is, the technology of creating God via the metamorphosis of reality.

I believe that there are at least two good reasons for "manipulation" of matter. 1) As discrete demiurge we would not allow a mountain of chipped away marble to surround a puny statuette, the ring of entropy (disorder) to surround a tiny bit of order that "has caused it". Therefore, for God's sake let's at least prophesize that the whole of cosmos will "order" itself into grace without ignoring any piece. 2) Space, time, mass-energy, our constituents if not our makers, are also our nemesis. In their presence, even a residual of them, full grace is not just improbable but impossible (see Science, Tao, Nirvana and the dualism they speak of). Only the total "manipulation" of matter into "spirit" will bring about grace, love, beauty in full.

Teilhard de Chardin could perhaps have proposed that independently from what the difference might be, love is the bond between esthetic and technology. Love unbound would see them coming together and coincide, into what? I propose into that which I called not Omega but Omega Seed. Then under the banner of love we would move on the track of technology and on the track of esthetic or, perhaps better put, with love as the indispensable splicing of desirable technology and bona fide esthetic, reality might slowly metamorphosize into Omega Seed.

If as the ever obliging Martian I look down (or up or sideways) upon the hominidi, what do I see: *homo faber* (technology), *homo scientificus, homo estheticus, homo theologicus* within the envelope of *homo amans* (loving). The manipulators are *homo faber* and *homo estheticus. Homo scientificus,* the seeker of "truth" as such, is always turned to the past (the only reality), while *homo theologicus* eternally fantasizing, is anticipating the loving one, God. If *homo scientificus* and *homo theologicus* get together, they can develop a technology of knowledge that is both past rooted and future oriented. A lot of love must shower on them to lubricate the cogs and keep to a minimum the heart of ignorance, intolerance, bigotry, self-deception and arrogance.

If *homo technologicus* and *homo estheticus* get together, the theory of knowledge developed by *homo scientificus* and *homo theologicus* would afford larger and larger chunks of the material world to vivify and become conscious of itself. The animation of matter would slowly extend the esthetogenetic process throughout the cosmos.

If there is such a thing as a priestly mode, it works as the simulation of perfection through the filter of human limitations, including mind. It plays such a game by the (rear) mirror technique. What it sees in front—the future—it postulates to be the reflection of what it thinks to be in the back—the past. That keeps the priest prisoner of an anti-future, anti-time paradigm. It makes him the paradox of life denying life, in the name of life: a nearsighted engagement of a clear sighted need.

If there is such a thing as the scientific mode, it works at the unveiling of truth as if truth were given (the priestly assumption);

it plays the game of the solicitor that does not identify in the solicited any part of itself (classical physics). The scientist observes reality from a self that—according to his "reality"—reality is least concerned with.

If there were such a thing as the technocratic mode, it would work at the manipulation of matter as if what matters were purely to make black that which is white, and to make white that which was black. In a better kind of wrongness binge, it would play the game of digestor, not much asked about the input, even less asked about the output (in cosmic terms) as long as it could sell at a profit.

If there is an esthetic mode it works on the creation of a world without residual means. A world that "at the end" is all sounds and songs and no instruments. It keeps the artist dwelling in deeper and deeper ambiguities that resolve the anguish of the moment by proposing a stronger one in its place.

Of the four, the technocratic mode is the most blind. It does what it does mostly for the wrong reasons (the consumerism-utopia), but it becomes in the process proficient in the magic of metamorphosis. This is not the only reason the esthetic finds the technological both appealing (indispensable) and impervious. It finds it difficult to pull it beyond the clinically functional. The esthetic is also aware that it itself, as well as the technologic, have propensities for extravagance which priests and scientists for their singular reasons abhor. The consequences of extravagance are idiosyncratic displays on one side—see the glut of narcissism of contemporary "art"—and the gadget syndrome on the other—see the shopping center.

For the artist the true vocation might be in seeking the universal that transfigures his personal suffering into the anguish of living and then extracts from it that which transforms it into the esthetic.

For the technocrat-technologist the true vocation could be found in the realization that matter is wanting in consciousness, sensitivity, spirit, and that to metamorphosize it into beauty it takes a technology of grace. The technologist must become a technologian, the theologian of reality.

Immersed as we all are in all these modes at different degrees of dedication and belief, we are, individually, torn within ourselves and also from each other. Anxiety makes us ready for indictment, if as priests we feel subservient to the scientist, if as scientists subservient to the technologist. . . .This conflict is not limited to the intelligentia because we all, uneducated, educated, scholars, look for the focus of things, for that which will not go away, for a *raison d'être*. The art of living, as far from narcissism as from masochism, escapes us in almost monotonous refusals. It probably escaped Teilhard also, afflicted as he must have been by the peculiar torment of seeing himself, and the best of himself at that, refused by Mother Church. But perhaps, fragment by fragment, he *lived* esthetics even more than being consciously attentive to matters esthetic. His work, more and more a cause for the church's refusal, more and more became the transcending of anguish, and as such a fragment of esthetogenesis.

To obviate the inequitable by way of piety and compassion is to act altruistically, not esthetically, for altruism lives only in the shadow of inequity. Once inequity is extinguished, altruism will be "out of work". Altruism can exist only if "evil" exists. The esthetic is something else, it does not exist as a function of something else. It is an end in itself. perhaps it is *the* end.

What of love? If the esthetic were a fish, water would be love, the indispensable medium. The water turns out to be for the sake of the fish, not the fish for the sake of the water.

Is there then an esthetic Teilhard? And how large is its shadow? What there is of Teilhard esthetics is to be found in the conversion—acceptable or not—of religion into esthetics. If religiousness is ultimately the longing for the beautiful, that which transcends the intrinsic inequity of the real, then Teilhard was eminently involved in things esthetic.

Chapter Eighteen

Monasticism and Reverence for Life in the City

Not attempting to define monasticism as such, I will limit myself to see if some of the premises of monasticism are radical enough to make it germane to more universal or diffuse phenomena. I will pick out what is perhaps common to all living reality. I call this common element the "Urban Effect".

To put it briefly, and quoting myself from my book *Fragments*, (Harper and Row) "The human habitat demands the coexistence and cooperative interaction (or synergy) of physical, physiological, and transphysiological reality (social, cultural, and religious factors). In this three-pronged reality, what works as the indispensable medium, the physical, also works as an unavoidable obstacle. Therefore, while consciousness and knowledge develop through the medium of physical reality, consciousness and knowledge are in turn vicitimized by its coarseness. In other words, the physical supplies the means to develop consciousness and knowledge while at the same time being 'noise' that impedes their growth. Thus, in an evolutionary process seeing the emergence of consciousness and knowledge, the space and time parameters are destined to 'optimize' their participation through progressive effacement, minimization, and miniaturization, consuming themselves and, in so doing, favoring transcendence into the transphysical. This process is the flowering and the triumph of complexity. See it in physiological evolution, see it groping now within the technological progression. A paradigm can be articulated: in any given system, the most alive quantum (a city, for instance) is also the most complex. In any given system, the most complex quantum (a city, for instance) is also the most miniaturized.

"The paradigm might dispose of the irreconcilability of substance and essence, matter-spirit, body-soul, because the second is generated from the first without this first being shipwrecked in the steady state of a fatal, unchangeable, solid, unknowable reality. Instead, what makes reality is the dynamic of a progressive vivification, of matter becoming spirit, the urbanization of the mass-energy, space-time original medium.

"The urban effect is the eschatological denouement of consciousness. The origin of the urban effect is in the ability that appears in matter, when matter attains certain morphological properties, to become responsive to stimulii and stresses in ways that are not the norm but the exception. By 'norm,' I mean that, given a certain condition (cause), there will normally be a certain response (effect)."

Throughout history, the city (villages, towns, habitats in general), the epitome of the urban effect, has been present and developing in parallel with the monastic habitat.

What is the fit between the monastery and the city? In a way, from the secular position one could call the monastery a protocity; from a religious position one could call the city a proto-monastery. The more accepted version is that the city is that which the monastery does not have or does not want to be and that the monastery is that which the city does without or is unable to produce.

I would go with the first proposition because at the end of the tunnel of time I see the *civitas Dei*, the monastery-city-God, that one could interpret as the fully defined gene of the cosmic process, the cosmo-genesis, the process we are part of and which is carried out by the persevering work of the urban effect. Then what could be the fit between monastery and city? Within the historical-evolutionary context one could divide the question in the three realms of past, present and future. Parenthetically, this distinction postulates a reality of time that is obfuscated and ultimately denied by what has been the animistic-revelational model of reality.

What has been the fit?

What is the fit?

What will be the fit?

Time-change is the milieu in which both occasions and their fit are laboring. Time-change are the milieu and the actors in one. In fact, since time is change, time is the *deus ex machina*. To time-change belongs not only the nature of the fit but also the constitution of both monastery and city. So quite suddenly the question becomes just about unmanageable—even for a mind with more abilities and knowledge than I have.

I will then, snail-like, retract my antennas and with reduced perception propose some points, hoping they might find the right context.

After a few sessions with Raimundo Panikkar and other brothers and sisters, I would not dare to attempt a definition of Monakos. I instead pronounce the name and some sort of image-evocation puffs out of my brain.

First, one thing seems to be certain. Monakos seeks community with others (brothers-sisters) and communion with the *causa prima* (which is for me the *causa ultima*—Omega Seed). If and when he/she does so, he/she is seeking the city, the urban condition (effect) and specifically the city of God.

Not to be cursed by arrogance and/or still prey to secularism, he/she speaks of an equitable society supported by effective and pleasing environs. Fair enough? No, not for Monakos. Monakos is inebriated by the divine spirit. Such spirit is in need of appropriate bottles. The awesome adobe (habitat) of Godliness.

Second, if the human phenomenon is divine (the adobe of the divine) and if the human phenomenon is complex (a metamorphosis of "simple" matter), then the divine is complex. What makes the divine operative, that is, what makes it tangibly transformatory, is the intensity, the concentration of its presence. It is the crowding of change into ever smaller and shorter time-space occasions. Crowding is typical of life, of habitat, of spirit. Crowding is the urban effect and it is spelled out within the paradigm of complexity-miniaturization-duration.

"In a semi-infinite cosmos, one could say that intelligence might be semi-infinite. That notwithstanding, the cosmos is pretty "unintelligent", not unlike the coldness of a water tank in which enough calories (energy) are stored to keep a water kettle boiling for days. Dilution breeds uneventfulness. For the quasi-unlimited expanse of the cosmos, the amount of possible intelligence dispersed in it is too small to reveal itself and act.

"Amounts of such intelligent stuff must congregate (urbanize) in particles dense enough to become agents of creation. That is the task of life and of intelligence. The mechanism through which the process is triggered and sustained is, evidently, miniaturization of the context.

"Finiteness can come to contain infinity through sufficient compression of events within its boundaries. Containment becomes an imperative, because noncontainment means the dilution of eventfulness down to the bare, squalid, randomness of elementary matter, the condition that life forbids to itself, because it stands for life's demise.

"According to the Big Bang model of the cosmos, we see the congregation-congestion of matter into a mega atom and, by the immense pressure of its denseness weighing on itself, the explosion of it into the expanding universe. In this process of "disintegration", matter defines finite focuses, making clusters of itself into galactic phenomena. But the cold, bleak uneventfulness of intergalactic space and the blind fury of the solar furnaces, the stars, lack intellection and are incapable of it.

"Intellection requires immense sophistication and involves unlimited and patient experimentation. Both need a medium in which the agitation level of particles is neither in the vicinity of absolute zero nor in proximity of nuclear fusion. The silence of the one and the shattering tumult of the other must be mediated by the gentle noises of moderation, the wind blowing, the rain falling, the atom tamed into urban effect, the synergetic associations of monads intensely pursuing the associative, constructive, integrative modes of consciousness.

"Complexity-miniaturization-duration locks matter to spirit with unbreakable bonds. The consumption of the first into the second is the only resolution. If it is genesis, the spirit prevails. If it is nemesis, matter prevails. Ultimately, it boils down to the fact that we can attribute more wisdom not to the father but to the son. Therefore, 'nature' the father-mother will have to give way to spirit, the son-daughter.

"If the universe finds the strength to reverse its expansion and eventually implode into the symmetric to the 'primeval atom ball', then the task of logos is simply to synchronize its operation on the universe in such a way as to maintain its own progressive metamorphosis (the progressive metamorphosis of it) without compromising 'today' what needs to be done tomorrow and keeping in mind that the last tomorrow would be the moment of total implosion. That means that such an accelerated contraction toward an ultimate atom would never occur. Along the predesigned implosion lasting billions of years, the spirit would nibble away from matter a sufficient and necessary amount to correspond to the design of the spirit consuming all of matter. Therefore, the final moment is reached not as an infinitely dense sphere of particles but as an infinitely miniaturized (point of) logos." (from *Fragments*, Harper and Row)

It is unfortunate that we obsessively speak of spirit and practically never try to see what is behind it or sustaining it. What is there is, in fact, nothing else but the urban effect of complexity, miniaturization, duration. Why?

If by magic we were to dispose of all the crowded occasions (urban effect) our planet has had in its 3.5 billions of years of organic life, the spirit of the earth would equal the spirit of Mars or Saturn: a spiritual whimper infinitely smaller than the spiritual whimper of the earth.

"Crowding is theological; congestion is pathological. To say that you or I do not endorse either paradigm is beside the point. Because we are ourselves "crowding" to the most intense degree and in the best sense of the term, and because we are also victims of congestion in the pathology of our environs, urban and suburban, it is necessary to understand the distinction, the chasm, between the two terms.

"Crowding is theological because 'God' is the epitome of the interdependence of the whole and the part to a (point-limit) degree where both crowd themselves into an inseparable, indeed, indistinguishable oneness. The all-inclusive theological reality, where particles suffer the presence of other particles to impinge positively on the health of each other—there is where the urban, theological, effect occurs.

"Such effect might be relatively simple as in bacteria or viruses, or it might be unbelievably complex, as in the human species. It might be a simple association of elementary organisms as the coral reef or a complex association of complex organisms in a (human) city, but the underlying nature of all these events is the same—the resonance (sufferance) of the single cell or individual with the whole and vice versa. That is the nature of crowding.

"The nature of congestion is described as that crowding phenomenon where confusion has taken the place of information, grossness the place of grace, complicatedness the place of complexity, waste the place of effectiveness, frustration the place of engrossment, squalor the place of fruitfulness, resentment and intolerance the place of sympathy and reverence.

"The pathology of crowding is congestive (descriptive of the gigantic city), and therefore congestion cannot exist if crowding is not there acting as the 'patient in danger of infection'. To dispose of congestion by the elimination of crowding is to eliminate the infection by disposing of the patient. Hardly a sane response to a difficult challenge." (from *Fragments*, Harper and Row)

Monakos, *per se* an urban effect of magnitude (all organism, enormously crowded with events are urban effects), cannot escape the rules of the spirit, the spirit which exudes from the existentialized complexity, miniaturization, duration.

In this light, it is tautological to say that Monakos and monastery are interactive with citizen and *civitas*). They are a manifestation of one and the same process, the crowding of events and consciousness in spaces-times that tend to shrink, to miniaturize and matter that tends to ephemeralize.

Therefore it is not only deceptive to think that city and monastery move on a collision course, but absurd to believe that one can exist without the other. After all the monastery is a habitat, a mini-city, and all things said and done, the city is that which instrumentalizes mind and sets favorable conditions for the spirit of wo-man to concretize.

But they differ; in fact, one can see the difference because they belong to the same mode: life responding and interacting with non-life. They are congruent and prone to adopt each other's strategies and tactics. But since they run on priority charts which differ, they are different.

From the position of the Monakos, the city is a monastery which has not successfully managed to saturate the habitat with spirit.

One could say that the monastery, from the position of secular wo-man, is the city that in calling matter upon itself and in etherealizing it into spirit has forgotten that reality is spirit bound in matter. Two different allocutions say the same thing differently because they are looking at reality from two different vantage points.

What seems to be needed at present is that city and monastery become conscious of the other's existence, of what each says and does. Then they might become complementary to one another, that is, reinforce each other's structure and deeds.

It could be instructive to develop two communities of about the same number of people within the same geo-political constraints, supporting themselves with similar activities, with similar technologies, one oriented toward secular goals (positivism?) the other oriented toward transecular goals (transcendence?) and find out where, how and when they "part company", by how much, if at all, they might differ and become confrontational. Such an experiment could add force to "escatological secularism"; it could weld the two because the (apparently) different goals might turn out to be the same dressed up in different garments or instrumentalized by different devices.

Escatological secularism (self-contradictory semantic?) could then be favorably assessed by both materialistic doctrines—Marxism and capitalism—once materialistic aims are upgraded from the level of false icons to the level of desirable instrumentations; that is, as the mind's intervention into matter in order to reorganize it into an extension of the mind's will, design and action—the action upon the disorder and the mindlessness of mineral reality.

All this will smack of arrogant secularism in deep ignorance of religious facts. It could well be so if not for another fact: such secularism might turn out to be more religious (reverential) than "religion itself".

Let's take, for instance, the question of equity (and suffering). For escatological secularism there is no hope for access to true equity and the remission of suffering, unless the whole cosmos becomes equitable: past, present, future; here, there and everywhere; an equity investing the subparticle and the galaxy. For them to become equitable, as well as for all else in between, means to metamorphosize into spirit. This is the tallest of orders and no religion I can think of stands by the "technology" necessary for its advent. I mean the technology of matter (time-space mass-energy) becoming pure spirit.

Then what becomes of reverence?

There is perhaps a *now* reverence and an *ever* reverence. The now reverence imperatively demands those deeds *now* that might be able to reinforce the blossoming of life and understanding among peoples and emanating from peoples toward all creatures of the planet (St. Francis, Albert Schweitzer, Mother Teresa, etc.). But there is also the *ever* reverence within whose arms the now reverence must find shelter to avoid futility, desperation and ultimately defeat.

The *ever reverence* cannot limit itself to an existential (now) imprint, but sees in such imprint a microscopic step into a reverential reality which by the fact of its non-existence is expectant of the process of creation that will cause it to come about. The coming about is pain-filled and immensely protracted in time-space for the reason given above; the imperative for all of the cosmos to sensitize itself and as such become the ever and total reverential witness of its own genesis, the genesis of the spirit.

In the city the *now reverence* is mandatory if the city is to serve well its own constituency. That means that both the words and deeds of the people must reverentially imprint the whole habitat. Such an environment will envelop the person and in particular will envelop the very young and the very old in tangible care and attention.

In the city the *ever reverence* is the powerful undercurrent of metamorphosis that the city itself causes and expresses. Such metamorphosis is documenting the process of a mineral reality reordering itself, via the urban effect, into organism. Such organicity is the self-organizing of the universe in the direction of mind-spirit.

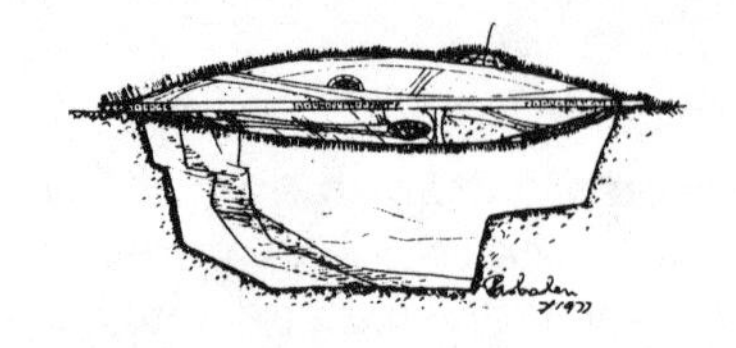

Chapter Nineteen

Spirit of the Earth

I would say that at least on one level the Arcosanti project is a contribution to the "how to" of planetization. Given an evolutionist's scenario, I see the enhancement of life and mind in it as inextricably bound to the urbanization of life and mind. Urbanization is crucially dependent on the same logistical effectiveness that governs any sort of organism. The imperative for this logistic is frugality of means in order to respond well to and to overcome the inertial mode lodged in the "physical" reality. The summation of all frugalities, the well-known do-more-with-less slogan, is in the achievements of miniaturization. Miniaturization is a clear, pervasive, constant characteristic of the living systems at the lower end of the evolutionary pyramid and ever more so as one moves up in it. With miniaturization, life succeeds in developing the complexity of its occasions. The two, miniaturization and complexity, mount in tandem. At the upper edge of this tide is the human organism. The modality that best fits the complexity-miniaturization paradigm is the "urban mode". It can be said that life is an urbanization process, a process that sees things and events coming together in subtle and highly complex, miniaturized ways and manifesting themselves as life, be it physiological, social, cultural, spiritual...

By discovering the urbanization mode as typical of all species and powerfully present within the organism of each species, (and flowing over into the urbanization of a specie—see the insect society, for instance) about 10,000 years ago wo-man invented the "city". Looking at the root causes of the Urban Effect, Arcosanti tries to "radicalize" the proposition by attempting to reinstate the working presence of complexity-miniaturization, quite probably in the best "Teilhardian tradition". Therefore, the "how-ness" of Arcosanti is the quest of a more fitting instrument for the human animal to go on in the transcendental quest for grace, also, I think, a Teilhardian aim.

Perhaps it would be wiser for me to stop here, but I will not. One can speak of how's when a kind of covenant is underwritten on the what's and the why's. Since I cannot sign the covenant as it stands now (at the present level), I also have reservations on the how's in a "pure Teilhardian tradition".

To be abrupt, what if the admirable and engrossing edifice of Teilhard's work were to be deprived of its foundation, God? Could it withstand the storm of mind? I think that not only it can, but it would turn out to be a far less ambiguous structure.

Then at the onset, I propose a God-deprived reality, that is, a different set of what's and why's, and then scout around for the appropriate how's.

I do not have the scholarly tools necessary to give full and persuasive arguments but that does not discourage me from guessing and hypothesizing. My hunches are that the very reason why we are clumsy at defining the how's à-la-Teilhard is because our what's and why's in the Teilhardian mode need revision. Revisionism speaks of disaffection, of ill-placed trust, even of treason. Be that as it may, my revisionism deals with a Teilhardian model that has been "robbed" of God. I propose to sketch and justify such "deprived" eschatology (the what's and the why's) and then take a second look into the how's.

Trusting that I have in me a Teilhardian imprint, I will not attempt to propose that my observations and reservations are apropos or well-targeted. I will simply and, I am sure, quite randomly go about my neo-Teilhardian eschatology.

To begin:

What of God in a reality that has "meaning"?

If God is not necessary, God is superfluous.
(If life is not necessary, life is superfluous.)

If God is predispositional (it predisposes), God is tyrannical.

If God is experimental (it experiments), God is fallible.

If God is creative (it creates), God is inadequate.

If God is playful (plays games), God is cruel.

If God and change (time) coexist, God is not the absolute.

If God, the absolute, is, change (time) is not.

If God is, then I, the discrete, am not.

If I am (the discrete), then God is not (the absolute).

Even if one were to succeed in chipping away at most of those propositions, the remaining fragments would suffice for the dismissal of an all-knowing, all-loving, all-graceful reality. It would turn out that reality, of which the hypothetical divinity is part, is a very wanting reality. The scenario seems to be then of a reality, its God or divinities necessarily included, in quest of fullness. This, by the way, sees Omega zipping away and for all practical purposes getting lost in a future too distant to be reasonably anticipated (prophesized). But then, eschatology does not deal with reasonableness or feasibility, but with desirability.

Since the absolute is by definition incorruptable, an eschatology that finds it necessary (desirable) to re-establish the absolute is, more than self-confounding, self-contradictory. By the same token, if reality is not absolute (as Tao is supposed to be?), divinity is absurd since one proposition excludes the other. As for the absoluteness of process, it would somehow seem the reverse presumption because process is metamorphosis of that which is into that which is not (yet). If it were not so, process would be at best revelation, the unveiling (to consciousness) of that which is. But since the act of unveiling per se would be process, the whole thing is self-contradictory: The unveiling of the absolute disposed of the very concept of absolute. The absolute to which the unveiling is added (the evolutionary process, for instance) is defrocking the absolute of its nature.

The *mysterium tremendum* persists and the notion of divinity is inadequate not only because if divinity were real it would be tremendously mysterious, that is, normatively ineffectual, but for the most compelling reason that if God turns out to be clumsy or cruel or hedonistic or infantile or hopelessly tangled in its own hubris, we may have to be careful how we deal with it.

I tend to believe that by inventing God, by adopting the revelational mode, we can unload upon God himself all the deficiencies besetting the human species. So perhaps in a gigantic Freudian slip our invention of God is our way of saying that if even God manages to mismanage its kingdom, who are we to do better? This escapism is as good as the one that says that the infinite love of God will ultimately see to it that the poor, wretched, living kingdom will find redemption and felicity.

But what if the postulate instead of being revelational were creational? What if this were truer: an infant self-creating reality beginning to see the possibility, dim and remote, of a process that co-involves each step of the process itself in the creation-construction of a graceful, beautiful, divine occasion? If it become plausible that evolution and God cannot coexist, but that instead evolution might be divinity in the making, it is up to our conscience to chose between the two and then begin to survey the equipment yards proposed by the chosen mode (eschatology) as to the equipment necessary, if not sufficient, for the task.

To believe and to propose that the how's applicable to the restoration of full grace, as purported by religion, can be the same as to the how's necessary for the creation of a hypothetical remote grace is unwise and dangerously self-deceiving. I would not care to be part of such deception.

Some points as to the two different eschatologies, revelational and creational:

Mode 1. Revelational

Given the postulate that reality IS and therefore the becoming is at most the unveiling (unfolding is the fashionable term) of it, it would seem that knowledge is all that is necessary. To improve knowledge entails the development of better instruments for knowledge. One improves oneself to become a better instrument for knowledge, a better mind and a better array of instruments supportive of mind's work. Ignoring for the moment the absurdity of the co-presence of the absolute and of an incrementally better instrumentation for the acknowledgement of it, one could propose that, instrumentation being for real, life and consciousness would be, at different levels of efficiency or efficacy, instruments or means to the end of an ultimately acknowledged glory: God. To be (and become) for the glory of God (a God that exists "in spite of" the means that are working at its acknowledgement). Since the goal is supreme, any means to that end are good means. (This would not be true if the end didn't exist yet because then the means would have to justify themselves.) The process is such that if there is deception in the end, the means, good or bad, are deception coded.

Here are some instances taken from the bins of goodness, within the revelational mode:

A. Since reality is benign (created by a loving force) reality is sacred (a manifestation of God). Any action that alters or attempts to alter it is disruptive, is destructive. That brands becoming with the indelible mark of an original, unredeemable sin. To become is to interfere in the grace of being. Any rhetoric that names becoming as "moving into grace" hides the fact that an action that intrudes into the absolute is an intruder in the absolute, period. White intrudes into black, black grays; to this perception of reality belong the outer fringe of the conservationist movement and the return to nature children.

B. Since the human phenomenon is God-created, there is an innate God-ness in *homo-primitivus* and his "natural" ways. Therefore, the tool-making of the early ages is okay. The music changes when the abstractional power of mind brings about the technological revolution. Innocence lost, Prometheus on the loose, nemesis befalls the species via technocracy. The idolatry of entropy (Second Law of Thermodynamics) proposes the rationale of a life as elementary as possible, protozoan life for instance; because it is less forcing an entropic reality to unwind itself (the converse for complexity growing in time and thus unwinding reality at a faster pace). Wo-man, the enormously complex, is consequently "the enemy". At the fringe of the appropriate technology proposing, or at least sympathizing with, the perception of an honorable stewardship of Mother Earth belong the "withdrawal from society peoples" posturing themselves "humbly and appropriately" for the best possible revelation of her (to themselves).

C. Since there is a God-given scale to things, it is a sin of the first magnitude to develop instruments or occasions "out of scale"—bad omen. At one end of the scale one wonders at the oppressive feeling a sub-atomic particle must suffer as it is subject to and coauthor of events (organisms) that are trillions or quadrillions of times bigger than its puny self. At the other end of the cosmic scale one wonders at the discomfort of a being that is by definition as large as anyone is able to even remotely imagine (God is all-comprehending).

To the perception that small is beautiful, belong (at the fringe) the zealots dedicated to the reduction(ism) of all things which "bigness" confounds. The fact might well be that scale and size are not at all congruous, compatible categories. Miniaturization expresses the gradient of complexity and not size. (Is the elephant in a bigger scale than the mouse?)

D. Then there is the true and proper idolatry of simplicity. The most direct way to respond to it might be to remind its advocates that there seems to be only one channel that can afford such a notion: the human cortex. That bewildering complexity plays itself out indefatigably instant after instant within a few ounces of flesh. All the intuitions, inventions, prophecies, creations originating from wo-man necessarily depend on this—insufficient—complex equipment lodged in the skull. If what we call simplicity is anything at all, it is the utmost complex synthesis coauthored by the cortex of the parameters on hand. That is the very opposite of the (entropic) notion of a pristine, innocent, simple, affective, wise reality. That is why holy evolution is so generously but sternly working at producing ever more complex futures so as to transform the simple non-living reality into a complex, live and vivifying process. That the most complex of all hypotheses, God, may get tangled in the advocacy and production of simplicity points at the absurdity of a model that proposes God and simplicity in one. Only a frightened, envious, greedy but cunning God would go for it.

Mode 2. Creational

The postulate that sees knowledge as the premise for the "transformation" of extant reality into future realities more and more radiant in consciousness and sensitivity (complexifying) means that *knowledge is not the uncoverer of grace, but is the instrumentation for the creation of it.* The difference is, at least in theory, immense. No less different ultimately must be the means, the how's engaged in the pursuance of two such different ends: revelation of grace or creation of grace. For instance, time is for the first one, at most, the surveying device of the extant landscape of reality. Time for the second is metamorphosis of the extant into that which is not (yet). If in reality the how's for the revelational and for the creational are not so actively different, it is because reality and life in it goes on creating itself no matter what the mind makes of it; and therefore the instrumentation willy-nilly follows suit. But how much more effective the how's would be if heart and mind were to be in the right place.

Running down four examples relating to the revelational hypothesis, one can point to the contrast between the how's of revelation and the how's of creation. (If there are two slogans grossly approximating the eschatological difference it is that the first pursues happiness, the second pursues worthiness.)

1. The understanding of an original reality that is not benevolent (revelation) but indifferent (evolution-creation) is only the first step toward a productive context whose task is to construct systems that more and more are capable of transforming "inert matter" into organisms or occasions radiant with spirit. Then the knowledge applied to good stewardship becomes the knowledge applied to metamorphosis with all the inseparable impending dangers.

There is another crucial reason for qualifying, if not disqualifying, a conservationism that swears on stewardship. It is that we all know that nothing which is here today will be here a million years from now, let alone a billion years hence. For a longer future even Mother Nature (earth) goes overboard. At the same time, a hypothesis that includes the possibility of a total recall, implicit in the creational postulate (see below) in the total oneness of the end, would have precisely the effect of preserving for that special moment at the end of time every single bit of the evolutionary process. But for such an hypothesis the how's are trauma-prone (metamorphosis) instead of maintenance oriented (stewardship).

2. If wo-man is not innately good but innately enterprising

(alive), his-her primitivism is not inherently ideal but just contextual. "Industry" is the way by which he-she tries to confirm and perpetuate his-her presence only inasmuch as that is the necessary premise in the use of the know-how directed at the transformation-metamorphosis of the context in which and of which he-she is.

For both (points 1 and 2) the food chain, comprehensive of all the living world, illustrates both the beautiful and bewildering miracle of life evolving and the deep, no less bewildering, inequity of its modes. Naturally, it takes the delicate interplays among the synapses of wo-man's brain to propose sufference as the pervasive aura of all-living and the inequity (a human category) that goes with it. Let's just say that fitness—survival-preservation—means I'll eat you before you can eat me (I'll outlast you).

It then would appear that there is a profound and troubling continuum of inequity in whatever becomes (life). To transcend and solve the good-evil dialectic might take no less than a total and immensely traumatic transformation-consumption of the known cosmos into a pure radiant gene, the cosmogene-sis of the universe, the creation of the Omega Seed; this, like any good seed, would be "resurrectional". This proposition is the reverse of the revelational proposition that sees the "more" of the beginning—God creating the cosmos—getting tangled into an entropic unwinding.

3. The notion of scale as an icon to being human is understandable but hardly meaningful unless one sees the whole context. I would propose that if the task is performed well (humanly), the system is right regardless of size. The ferry does better than the two-seater boat, the power loom better than the hand loom, the city organism better than an aggregation of houses, and so on.

Underlying the slogan "small is beautiful" are many not so beautiful worries. If "it" stands for the not small, some of them would go as follows: not being up to "it", incapacity of being casual in the making of "it", being afraid of "it". "It" being the unknown, the unproven, the incomprehensible. But if the hypothesis is creational, the instrumentation has to be in scale to the wisdom, the compassion and the anticipation of a process whose pedestal is maintenance but whose mode is transcendence.

4. If reality is originating an aim out of its aimless beginning, and if such aim has to place its hopes in a complexifying-

miniaturizing process extending "forever" in the future, the instrumentality that "can do the job" must reflect and must be capable of dealing with such progression. Therefore, simplicity is not only immanently dull, but it is also eschatologically evil. It is an obstacle to transcendence. It is entropy. Entropy mandates simplicity so as to postpone indefinitely the onset of entropic death. Unfortunately this says that a non-living reality is less entropic than a living one. It is also the death wish of a reality that having had a taste of life, says "no" to it.

What then differentiates the behavior of a society pursuing the revelational understanding of all that is, God, and a society partaking in the creation of that which might become God? Naturally societies do not work explicitly toward either aims because the components (us) are engaged in a mosaic of activities confusedly bunched and labeled: ego trips, social aims, political orientations, economic goals, national interests, cast safeguards, ethnic preservation, religious orthodoxy, etc. The fact remains that the animal *in toto* expresses propensities and prophesizes future landscapes, while the grinding of the components (us) goes on with more or less regard to minimal levels of decency, equity, compassion, reverence. The historical perspective helps to clarify the orientation of a society. Often in words, sometimes in facts, the revelational or the creational bias can be helped by the historical perspective.

For both the revelational and the creational hypothesis the professed aim is Godness, but given the different eschatological lights, the first one will tend to abide by the Christian rules of altruism, piety, love and equity; the second can only see those rules as necessary but insufficient conditions for creation and for genesis for two reasons:

1. Equity and love pursued in revelational terms are illusory. The food chain, for one, stands in the way. Incantation hymns aside, the living reality has to face its own unescapable "violent" nature which is the very nature of organic reality. Wo-man carried into the human kingdom the same condition, embellished with the typically human additives: greed, fear, hypocrisy, ambition, cruelty, etc. That is, the limited goal of revelationism is out of reach because of the very wanting nature of the "experimenter" (food chain).

2. A condition of equity achieved through stewardship and by abatement of greed, fear, hypocrisy, ambition, cruelty, even if possible, would leave out the bulk of reality; it would leave out all of the past (debased therefore into means to an end); it would leave out all of the cosmos not redeemed by the knowing of God; and it would leave out all of the future, separated from those ills only by the brittle, tenuous diaphragm of a moment, the present, of (pseudo) equity. Then it might appear as unavoidable that it is only by the demise of cosmic parochialism and the rejection of a dim-witted notion that sanctity is just around the corner (that equity is achievable with the tools on hand, stewardship), that there is a leverage point making possible, if remote, the chance of reality to become equitable and divine.

The revelational mode puts up—and rightly so—countless red lights warning of impending disaster. The creational mode has to be serious in investigating the changes and act consistently with the findings, but it also has to be true to its own imperative: the creation of the, as yet, non-existent. A non-existent that is going to be unimaginably different from the present for the same logic that makes the present unimaginably different from the past, the ten billions of years of uninterrupted metamorphosis.

Then the "how-to" manual for a responsible present is to pick up the pieces of a still alive composite of civilizations, reorder them and in so doing transform them into more intense, more alive (more miniaturized), more interactive learning and performing enclaves. It is an active, willful commitment to the urban effect and what it means in evolutionary creational terms.

It is not a commitment to a religion, but instead a religious commitment. If religion means binding, it is like saying that there would be binding, "religiousness" but without binders (religion). Binding would then be that which in the face of harsh, indifferent premises, the "non-living" and the necessary violence and inequity-prone process to evolution, encourages and confers staying power on the cooperative, interactive, social, cultural, reverential, frugal modes of communities giving them a chance to develop within a context—at least for now—of limited resources and energy: limited space, limited time, limited intellect, limited resourcefulness, limited altruism, limited imagination, limited knowledge, limited wisdom, limited compassion, limited love, limited grace. But I add once more that all such limitations compose a barrier to be broken eventually so as to reach and create the unlimited real beauty of Omega Seed.

If on the other hand all that is demanded is an intolerant or lukewarm endorsement of a religious dogma, or the dogma of stewardship, the search for and the application of the how's will not be much of anything. The malaise will not subside and the ego will eventually have the field to itself. Since there is nothing lukewarm about ego, its modes are then assured victory, the victory of intolerance and of self-deception.

Without passion in what we do, the future is gray at best. But if the ego is the passion, the future is full of dread, dread-full. To step out of the ego (Buddhism?) without leaving out the passion (Christianity?) is the "trick" that only a trust in something that transcends the values of the ego can perform. What I find unacceptable in the revelational mode is that since divinity, which transcends the ego, has preceded and will outlast it, the ego finds great excuses—if not a lot of comfort—in ignoring the whole issue (rendered mute by revelation) and go about its own not-so-merry way.

In addition, if I see in God the Almighty, my love for "it" must survive a break through the massive barricades of suffering (eons long). In His might He may well ignore suffering; my puniness cannot, teased also as I am into the hypothesis that He might just be kidding around.

And then if by chance I see in God a groping reality, I give to it a date and birthplace that coincide with the date and birthplace of the cosmos. All afflictions of the cosmos are God's affliction. All wants of the cosmos are God's wants and defects, and inequities. That is, "God" is the cosmos. "God" is in the process of genesis, the genesis of Omega Seed (the creational hypothesis). If so, the Arcosanti project finds a fit within the evolutionary mosaic of how's because, as stated before, it endorses the creational hypothesis of an Omega Seed by way of a process-creation (evolution) working toward the infinite (grace-full) complexity of a reality metamorphosing into spirit. As a speck among other myriads of specks, Arcosanti wants to tease grace out of an immensely "unconcerned" reality, the mass-energy-time-space cosmos. This is a process that I think would find Teilhard de Chardin sympathetically if not wholeheartedly on its side.

Chapter Twenty

Techno/logy—Theo/logy

Things are not well in the world. This is as much the fault of our secularity as it is of our religiosity. And it is not true that if secularity weighs 1000, religiosity weighs 1, therefore we can ignore it. Most of what we do as secular wo-man we "perpetrate" as religious beings. That is, we barter part of our soul whenever we barter our goods, including our own work. That compels us to put religion at least on a par with secularism. I hasten to say that I believe that religion is failing wo-man; a sin worse than secularity failing wo-man.

Monasticism (not always religious) is in many ways at the core of religious communities. As such, it embodies the sharpest features of such communities. It has radical voices speaking for it. Its radicalness also explains its historical rootedness. Tradition, self-reliance, self-discipline, mental activities, are consequences or attendant characteristics. But two main aspects are at least presumed for all monastic communities:

The trust in a transcendental reality and a frugal praxis; that is, the concentrated attention on two main opposites. On the one hand, an interest for "that which is not of this world", that is to say, not limited to the perceptible. And on the other hand, a quasi obsessive recognition and attention to things of the body; the measuring, considerate and stringent, of all exchanges and relationships between the I and the other-than-the-I. Frugality, in other words.

Therefrom is the irresistible appeal monasticism shall have for the project. Presumptuous but true, as far as I am concerned, is that we can do for monasticism at least as much as monasticism can do for us. From a recent conference, it appears to me that we might well embody the monasticism of the present, that is of a reality which is beginning to perceive the indissoluble, concrete anchorage and generation of the spirit by and in "matter".

A short personal disquisition: Raimundo Pannikar, a religious scholar, presented at a conference his view on monasticism, part of which included the guerilla monakos. From his description, I found enough to identify myself with it. Here is why. In the late 30's I rebelled against social hypocrisy; in the 40's against military induction and then against neo-fascism; in the late 40's I rebelled against my profession; in the 50's against ecological mindlessness and against social-cultural materialism; in the 60's I rebelled against urban barbarism; in the 70's against what I believe to be religious bigotry. It has been a cumulative rebellion in the form of an increasing commitment and action. I am more than ever at the barricade, but as in the preceding 40 years, not to defeat the adversaries but to offer alternatives. Now I am trying to build a gun which might be able to shoot bullets of coherence, tolerance, dignity, reverence, and why not salvos of transcendence? I am on the look-out for other guerillas—monakos and specifically a mix of artisan-technician-organizer-holy wo-man to be a cofounder of the Arcosanti Monastery No. 1.

End of statement.

Is there a soul distinct from a "techne" of soul making? If the answer is yes, theology is unhinged from technology. If the answer is no, we had better resign ourselves to the reality of an unbreakable bond between what techne does and what soul becomes.

I am throwing myself to the lions, so to speak, because I pick up what many minds see as the devil incarnated, technology, and splice it to the spirit and call the splicing indispensable, in fact providential. I do, in fact, more than that. I say: *The technology of matter becoming spirit is where reality operates.*

Does it matter that *matter* has to exist and that matter has to undergo "technical" transformations? It evidently does. Without:

The presence of *being*, that is, *matter* yet to be "immersed" into time-space and without:

The transformation of such matter by its "immersion" into time-space process nothing *is*.

A nothing spirit is then the consequence of the non-being and the consequent non-becoming (non-process).

For eons matter, via process-metamorphosis, has managed to make "soul" by way of a techne of the flesh. Of late sticks and stones were picked up by the flesh in a bold technological move and technology itself took a sharp turn to, but also away from, flesh as such. Ever since, two million years or so ago, soul and techne have made, to many minds, a faustian covenant and we are for those minds, the marked children of it.

To my mind the above is an indispensable premise to any doing of wo-man, Arcosanti included. The premise that *homo faber* is also a symbol of *cosmos faber*, the "fabricating" the cosmos is engaged in, the cosmo-genesis that providentially is a soul genesis. The genesis of spirit is is not of spirit (God) generating matter with all the ultimately inconsequential history of a dispensable process of becoming; it is instead matter generating spirit within the excruciatingly noble and suffering process of metamorphosis.

Therefore the urgency of *doing as thinking*. Because the thinking per se—and I include in it contemplating-meditating—is the beautiful flower of a *doing* that has generated brain-mind, that "thinking machine-performance" of immense nobility and resourcefulness at the apex of the pyramid (hierarchy) of the food chain. Without such a monumental pyramid, thinking-meditating-transcending is impossible, in fact inconceivable.

There was no thinking when there was no cortex to think with. *The thinking cortex is thought.*

One of the paradoxes of spirituality is that it is bound to (generated by) materiality and there, at the critical hinge of matter converting into spirit is Monakos.

We lay people go about the business of survival and affirmation within a cloud of praticality inflated with spiritual longings.

Monakos, at least *the* Monakos, goes about the business of reality with a firm grip, at times a self-destructive vise, on body and soul. It is a quickening of the body in direct touch with the raw physical-physiological make-do; a quickening of the soul that is none other than the quickened body transcending the physiological splendor it embodies, the uninterrupted miracle of the self-transcending food chain.

In this context of a spirituality locked in the sweat of the flesh, the techno-logy of evolution, I am going to elaborate on the environmental-architectural (arc-ology) work leading to Arcosanti and beyond.

What follows is a commentary on over 200 slides illustrating the process I have been engaged in for over 20 years.

I have to try to explain why we are doing what we are doing and how we are doing it. To begin with there are two resources that are coauthoring present reality (solar system): the physical sun and the spiritual sun. The physical sun, as far as I can see, is the originator, or the father, at least in the solar system; the offspring is the biological life culminating in the human and spiritual presence. Those are two radiances which have to be present in what I see is the habitat and the habitat is represented in this instance by arcology, architecture and ecology:

In order to try to explain the idea behind arcology, I have to present my view of reality, or a hypothesis of what reality might be. I start with a quite normal scientific notion of what reality might be. One premise is that I believe in time as reality that comes about. Time is change. We have an origin which is mysterious, and from such an origin, we have a genesis of the cosmos, a mass/energy space time universe (cosmos)-genesis. I call it the universe of indifference, or the deterministic way, relatively speaking. It is insensitive, structural, simulable, predictable, granular, rational, statistical, entropic and segregational. Those characteristics somehow are appearing again and again when you look at the city mode in its least valid attributes. This genesis is about 10 or 15 billion years in development. Then, within that context, at least in one spot, but possibly in billions of spots throughout the universe, appears a genetic way, the biological universe, the bio-genesis, the universe of innocence (before the appearance of conscience somehow the idea of good and evil seems to be absent). Such biological genesis is social, cooperative, sensitized, non-simulable, irreversible, selective, evolutionary, durational, unconscious and instinctive. Within this process that on the earth seems to have been going on for about 3½ billion years appears the cultural way, the human universe, the homo-genesis, the universe of conscience. It is cultural, social, cooperative, compassionate, not simulable, willful, irreversible, evolutionary, durational, conscious and mental. And this has been about two or three million years in the making. Then I do my own extrapolation; I suggest to myself that if this process of the creation of conscience and spirituality goes on we might reach the sacramental way, the theogenesis, the divine universe. It would be seminal, aesthetic, resurrectional, totalizing, loving, irreversible, non-simulable, genentropic. In this process, again as I see it, you move from a relatively simple kind of phenomenon, a mineral reality, the mass/energy, time/space universe, into more and more complex, more organized, more sensitized kinds of events and phenomena. So much so that you could say that there is an escalation of cooperation, compassion, liveliness, intensity, interdependence, association, innerness, sensitivity, response, self-reliance, learning, memory, hierarchy, knowledge, passion, anticipation, care, transcendence, reverence, etc. And it all seems to be pointing again at an exponential increasing of complexity. Now, given the nature of the physical reality, complexity can only be possible if it's associated with the parameter of what I call miniaturization: more being able to be present in less, more organization, more interaction, more acknowledgment, more memory, more interdependence, more cooperation, and so on. So in this movement from in-indifference into innocence into conscience into transcendence, I see the origin of what I call the paradigm of complexity and miniaturization. To try to somehow explain a little more what I mean by complexity and miniaturization, I use the example of a seed which is planted in the ground and becomes a tree. It is not so much an explosion of a small thing like the seed into a large thing like the tree, but it is the implosion of environmental elements which are relatively indifferent to each other into what the seed organizes them, through the genetic matrix. From the ground, from distances which are relatively small you get minerals and water; but in the air you are dealing with a cosmic process. The energy from the sun comes from millions of kilometers away and it's through this coming in and being captured and transformed and transfigured in many ways that you have a developing organism which is inner-motivated, inner-driven, highly organized, hierarchical, full of cooperative elements, full of interactions, and naturally a miniaturization in many ways of the scattered and relatively indifferent container which is the cosmos. This is my brief explanation of what complexity and miniaturization are. I see it as valid for bacteria as for any kind of creature which develops through the evolutionary process—for the human, and also for the association of humans.

That means it is valid also for the city and whatever might be coming after the city. In fact, as a larger extrapolation, I feel that perhaps the cosmos is seeking its own genetic matrix, and I call that the "Omega Seed". As I see it, it is at the end of a process of genesis, not at the beginning of it. In other words, the cosmos doesn't have a genetic pattern, *it has to create its own genetic pattern.* Such genetic pattern would confirm my paradigm. *In any given system, the most complex quantum is also the liveliest; in any given system the liveliest quantum is also the most miniaturized.*

In the Eastern mode there's a tendency to develop very complex mental and spiritual systems with a certain indifference toward the physical parameter, which is the miniaturization parameter. So one has what I tend to call the idolatry of the spirit. In Western society, it is the opposite: there is a certain facility to ignore the spiritual and the non-tangible aspect of reality. This includes ignoring the complexity of reality, especially in contemporary terms, and a tremendous ability to work with the physicality of reality. So miniaturization becomes a paramount and extremely successful process (the electronic revolution is a good example). But such emphasis is somehow at the expense of the truly mental and complex aspect. So I feel that if we "East and West" could come together, we might be able to develop on a temporal axis a more harmonious and more successful, and ultimately more engrossing kind, of reality. What I've been trying to explain is represented, at least in my mind, by a model which I call the evolutionary model, or the creational model; I mean the same thing by saying evolutionary and creational; I believe that evolution is a continuous process of creation. Through the filter of the present, reality (which is the past), is engrossing, is becoming more and more "something". In this model there is then an "empty socket" representing that which is not, the future. The future doesn't exist, it has to be created. The process of evolution, at this point, seems to be pointing at the capacity, via cortex, of developing into the empty socket a willful and possibly directed concern. I see the concern to be able to develop and to create grace, beauty, equity, etc. My problem with the traditional model of reality is that in it I'm not presented with this capacity for creation so much as I'm presented with the openness to a revelation. For this revelational model, reality is a phenomenon that tries to understand what truth is, tries to understand what goodness is, tries to understand what divinity is, and so on. In it wo-man is trying to be revealed or to reveal to itself what reality is through the help of others and through other institutions to understand and by understanding, to return to a grace which is pre-existent to the genesis of the cosmos. This revelational model relates to a kind of animistic universe in contraposition to the evolutionary model which relates to a non-animistic universe. I'm somehow, rightly or wrongly, working with the creational model in mind, in order to step closer to the human condition, but without losing sight of the process of complexification and miniaturization, underpinning the creational model—not to be confused with the creativism/creationist model which is animistic to the bone—I propose that we keep that in mind when we design our habitats. Such habitats are very important because we not only make environments, but we are made by them. If we do so I think we will be in a better position to solve some of the very crucial problems that we are presented with. I list some of them here: the alleviation of the ecological crisis; the self-responsible use of land, air and water; the reduction of the pollution caused by technological society; the perception of the nature of waste, affluence and opulence; a reasonable sheltering of man on his way to nine or more billion individuals; the resolution of the problem of energy depletion, distribution and consumption; the resolution of the problem of segregation of people, things and activities; the responding to an increasing encroachment of remoteness by way of gigantic services and the bureaucratic machines connected with them; and then, maybe a better trust in ourselves and the future.

Leaving out the first generation arcologies (see *City in the Image of Man*) I will now say something about the second generation arcologies and move into the use of energy. To this end I will illustrate very quickly a number of effects:

The *greenhouse effect* is a membrane that seals off an area of ground that can be cultivated, extending the growing season to practically twelve months, and also saves a great amount of water. This is very crucial if you realize that the problems we are facing now are, among others, our food resources and the encroachment of arid lands all over the globe. With the "greenhouse", one has intensive agriculture, limited use of water and extension of the seasonal cycles. This is the *horticultural effect*. Then there is the *apse effect*.

Some structures can take in the benign radiations of the sun in the winter months, and tend to cut off the harsh radiation of the sun in the summer. By the *chimney effect*, which is connected with the greenhouse effect, one can convey, passively, energy through the movement of air; the heat from one area to another, and I'll show you later what that means. So we have those four effects; there is also the capacity of masonry to accumulate and store energy—the "*heat sink*" effect. With relatively large masonry, one can store energy during the warm hours of the day, and give it out during the cool or cold hours of the night. The intent is to see if these five effects can be organized around what I call the *urban effect*.

The urban effect is the capacity of mineral matter, to become lively, sensitive, responsive, memorizing. A bacteria at the very beginning of the evolutionary process is an urban effect. Particles that were sitting, let's say, idle and indifferent began to interact, they began to acknowledge each other, and they began to talk to each other. This transformation of inert into living stands for me as the beginning of what I call the urban effect. I justify that, at least partially, by listing here some of the characteristics of what the urban effect ideally would be: synergy, intensity, consciousness, transcendence, frugality, sensitivity, beauty, knowledge, interdependence, coordination, cooperation, compassion, and so on. I take this processing of the inert into the lively as a universal process, at least on the earth, through the evolutionary escalation, and I call that process the urban effect. If we were able to coordinate those six effects together, then we definitely could save on resources like land, water, time, energy, minerals, and have a better ecological sanity. We could cut down on pollution, segregation, waste, bigness, bureaucracy, isolation, and alienation. And we would perhaps be in a better position for learning, for integrity, for health, for reach, for identification, for cooperation. The next step is from the conceptual or the theoretical, in terms of approach, into what the physical might become. If the greenhouse is on a flat area, quite definitely you need energy and equipment to ventilate the greenhouse in the summer, but if you slope the greenhouse, you introduce the chimney effect, and the greenhouse is self-ventilating. One can take the greenhouse and make it into a wafer, a sun collector, then you can incorporate the sun collector in the roof. This is the normal way of going about producing hot water. Or one can make the greenhouse with two effects: one as a sun collector, and the other as a food producer. Warm air can be pumped into the house and can produce some vegetables, as a form of energy. If one slopes the greenhouse, one can convey the energy which is, let's say, the warm air, automatically through the chimney effect. If one takes the house and transforms it into a multiple kind of aggregate like a village or a town, and then one enlarges the greenhouse proprotionately, one begins to see the possibility of having a solar town. But maybe one has to transform the morphology of the individual houses and make them into a masonry that produces what I call the apse effect and the heat sink effect. Then one has the six effects coming together.

Arcosanti is an ongoing process pursuing the urban effect within the context of the other five effects.

To close this loose circle around the reality of spiritual wo-man, the archetypal monk-nun struggling to emerge from *homo faber*, this techno-logian, carrying on the metamorphosis of matter into sur-matter, I want to emphasize again the crucial interconnection of matter, body, mind, soul. That is the imperative or what I call the urban effect. The urban effect is intent at the extrusion of meaning, lasting meaning, out of a mineral (mass-energy) reality, powerfully but not fatally swaying the conscience generating within it. By way of such genesis, which I try to identify as a techno-genesis capable of a metamorphosis of radical nature and immense dimension (a cosmic metamorphosis), the fruition would be the seed of the cosmos, the Omega Seed and within it the resurrection/ manifestation of all and everything. *That is the end of time and the full dominion of duration.*

Chapter Twenty-One

Love

Love, like any difficult "notion", can be better circumscribed than defined. That is, a description of love might end up in a crisscrossing of definitions and labels that reveal at the center a big question mark: love as such. That means that love is better referred to by listing what is not love than attempting to define what it is. Is love attraction or interference? Is love for what we are, possess, do, or is it for what we are not, possess not, do not? Should love for the beloved for what she-he is not but could be, has not but could have, do not, but could do? Is love an unremitting conflict with the "evil" which is and for the graceful which is not but ought to be in ourselves as well as in others? If my love for the executioner were to be strictly for the executioner in him, then I would love evil; that is, my love would be the denial of love. It would be hatred, the hatred that makes for execution. My love for the executioner is what is left out of the executioner mold and what can be put to work in the direction of what ought to be the opposite of the executioner, the opposite of hatred.

In reading Janis A. Rose's *Mobilization of Constructive Human Characteristics*, I find myself in agreement if "love" can be qualified in some way. The qualifications I am offering here may be somewhat contrasting with some of the "qualifications" implied in the paper.

Perhaps love is the rejection of insignificance (that which has no meaning) in things and beings and perhaps there are four main "ways of love". Four ways we find to connect with what is outside of ourselves, and gives it meaning.

1. We are possessed by an exuberant, excruciating, irresistible effusion of care for that which is the context of our reality—gods and people included. St. Francis may belong to this way, the mystical way, the way of oneness.
2. We are clearly acknowledging the precariousness of wo-man's condition and the fundamental inequity of reality. Our compassion toward the suffering is expressed by our care for reality.
3. The third kind of love is the love for truth of the scientist-philosopher, vacillating between the mystical and the esthetic but still forcefully loyal to a notion of objectivity.
4. The fourth kind of love, eminently mind-full (intellectually and emotionally) is the love for beauty and perfection, the esthetic way, the creative way. The artist might belong to this way.

Of course, there is the passionate love of the lovers. In it are mixed and "exasperated" the four kinds, and the explosive mix is best described by poetry.

The first kind is instinctual (genetic?) naive, optimistic. It tends to be boundless. The second kind is experiential, cultural, suffered. It is clearly contextual. The third kind is "objective", analytical but pursuing a universal structure comprehensive and explanatory. The fourth kind acts immanently and may succeed in producing "timelessness", the esthetic.

The first has no enemies to love because it has no enemies.

The second has enemies and "loves" them.

The third kind has one enemy, ignorance, and it is out to conquer it.

The fourth has enemies, mediocrity for instance, and battles them.

In the first kind is found the infant saint.

In the second kind is found the martyred altruist.

In the third kind is found the uncompromising thinker.

In the fourth kind is found the "Artiste Maudit".

When the loving-lovely person of the first kind loses anima, infantilism and dementia might set in.

When the tragic hero of the second kind loses anima the spin-off is melodrama.

When the seeker of truth of the third kind goes astray, his paradigms gather dust.

When the writer of tragedy of the fourth kind loses anima, the consequence is sterility, squalor or even cruelty. We all muddle through life loving in the four ways, now innocently, now suffering, now illuminated, now creating, or creating innocently, innocently suffering, knowingly suffering, creating in sufferance, etc.

What I fear and dread in the contemporary scene is the cheapening of love, through the reinforcement of sexual immaturity and promiscuity as well as in other ways. Given the relative coarseness of reality and the pale, miniscule presence of life in it, we might lose sight that love is a fire, a fire of innocence, a fire of sufferance, a fire of inquiry, a fire of creativity, or is it a semantic beggar?

Circumstances presently propose to society—the consumerist person—not fire but electric blankets.

One reason for this cheapening is that the fire of love is or can be uncharitable, even if driven by true compassion, when the uncharitable fire of love is at odds with consumerism, the electric blanket syndrome.

The dilemma is truly monumental. Can we truly love in comfort and security? Perhaps not, if for no other reason than electric blankets are rationed to a yet small minority.

It is not disparaging meekness and gentleness to propose that they both can be learned and carried on if the will to do so is there. As for love, one might grow into it more than one can learn it.

But that same reality (relatively mindless and indifferent) which might make love uncharitable and any hope for a near future of equity quite certainly illusory, makes for a limited hope for transcendence but only if we stay clear of "love now at any cost". That is not procrastination but realism.

Realism and compassion ask from us the acknowledgement that equity is indivisible and lasting or it is not (self-deception aside). Furthermore, any equity that is not retroactive is inequitable inasmuch as it is founded on non-equity, founded therefore, on the unenviable condition of inescapably falling short of the target. This is the crucial reason why love must be a fire, the burning fire of transcendence since it cannot be the fire of justice. This, if nothing else, is that which makes inescapable the tragic-beautiful nature of life with the love imperative in the eye of it.

But the love imperative has a physical component which is as mandatory as it is the physiological process which makes the reality of mind and passion. The physical component is the instrumentality of civility and culture. Only an animistic reality whose paradigm is that spirit precedes performance can dismiss the context and the necessity of such instrumentality. (Angels need no bathrooms nor libraries.) It is unfortunate that for such paradigms the "supremacy of spirit" goes for the dismissal of that which quite possibly has invented it, the wo-man mind and all the "plumbing" that comes with it.

That they, spirit and peformance, (the evolving cosmos) may have come together or that they are one and the same helps to sink one step further in the obliteration of time-change-meaning-equity-congruence, love. Anything goes in the circus where past is present, future is *déjà vu*, revelation is fulfillment, the maggot and the maiden are interchangeable.

Love is at the opposite pole of pantheistic-Nirvana scenarios, the scenario of self-deception. Love is ultimately the metamorphosis of stone into spirit, that which stone is deprived of, the whole of the known and unknown cosmos, of all stones that is, if equity has to prevail. God, oh God, how much stone is there?

Equity is not at the end of the stony tunnel of "matter" (mass-energy, space-time), but it is where and when the consumption of the tunnel is achieved and total, therefore, it is unending task of love, the transformer, transfigurer, transcender of reality as we know it or as we conjecture it. Whatever it might be, it is suffering (see the food chain); therefore, it needs trans-figuration.

Then existential love cannot be more than a provisional effigy of its true self. With the fixation of the effigy, idol-ization of it, idolatry is ready to lift its intolerant reptilian head for the sake of the fitting ideology—Marxism, capitalism, conservationism, simplicity, technocracy, monotheism. . .Then compassion becomes charity and philanthropy. The giving by the chosen if and when they so choose toward whom they choose.

What of the love affair between wo-man and nature? How much of it is in the eye of the beholder?

A bacteria has no pangs of conscience about nourishing itself at our body's expense, almost, if not just as, a precariously balanced brick has none in falling on one of our skulls, nor has a frigid wind in killing the unfortunate. Loving Mother Nature is a relatively useful euphemism for the irreducible fact that we are and we become because she exists (and becomes), but as far as we can see and test, we are her unwanted children. The moon or any of celestial bodies belonging to the solar system are as far as we can prove the norm of Mother Nature. The terrestrial biota is, if not the only, certainly, a rare folly of hers.

It is not up to her, immensely overpowering as she is, to tell us what next since it might well be "in the cards" that our preposterous task is to transform her in our image (life's image).

To those who turn their noses away from such demonstration of Western arrogance, I offer a sledgehammer and stand aside to watch in what order (under penalty of death) or delay (revulsion) they will smash a stone, a termite nest, a puppy, a baby. The human condition is built on similar propositons. That puts in a vernacular context the focus on the dilemma of Mother Nature, the food chain and equity. It also points clearly to the act that Love (with a capital L) is a futuristic propensity more than an available or achievable condition.

Odd as it might be, this way disposed of endless rhetoric and mindless slogans like: everything is sacred, we are all brothers, nature loves us, He, up there, cares for us, it is all for the better. . .

False love is an insidious sickness. It kills by all available proxies, the most hypnotic included, the beautiful landscape, the (rabid) bunny, the enchanted forest, etc.

False love is a misconstructed relationship. The landscape is beautiful, the bunny is sweet, the forest is enchanted, but there is an indifference and a ruthlessness intrinsic to them. That must not escape intellection, if intellection is after a fair assessment of our perceptions. If the ecosystem is an expression of love, it is against the "better judgement" of its component. The fire of love is in the bitter recognition of that and of those acts that take it into account while pursuing an eco-system which might be even more astonishing. A future less astonishing than the present is a lesser future which is why love must go for a more astonishing future. Therefore, the fiery nature of love.

Not to defeat its pursuit one must discourage a simplistic and deceptive running after what is not in our means to reach. To engage in the "ways and power of love" is of the essence for wo-man provided we are aware of the distinction between sweet abandon which has its legitimate moments and what is called for by the Tiger Paradox-Paradigm, as I label it, the beautiful-inequitable ways of the food chain (evolutionary) reality. To this end we must also understand that we, the living (noosphere) are and will be part of its gargantuan appetite as long as we have vestiges of physiology in our make-up, therefore, again the necessary fire of love given to the transfiguration of the beast.

It is in the nature of love to break bondages as much as it is to bind together. Then the act of love is sufferance even for the innocence of number one.

To suffer the experience of living is love.

There is the lamentation and the resignation of the innocent awed by the surprise of suffering.

There is the indignation and the indictment of the martyred caught in the web of things human and inhuman.

There is the intellectual anguish of the mind deprived of knowledge.

Or there is the "detachment" and despair of the creator segregated by force of circumstances (human and in-human) from the vision (action).

As it is for most fields of endeavour, ultimately and often cruelly, the proof of love is in the lovely pudding. Without the "product" the process is problematic, too ambigious for its own sake, too tentative, and now and then, just hypocritical or irresponsible. The love of the prophet, for instance, belongs to a pudding to come. It will come, it could come, tomorrow or some eons hence or at the end of time or never.

What distinguishes the love "esthetic" from the love "divine", is that the first is immanent, concrete though fragmentary (works of art). The second is anticipatory and simulative. It "has that" which is not, by telling himself that that which is not, is.

The "altruism of love" must eventually come to grasp the complexities of reality and acknowledge the danger of "ignorant love" so as not to blind itself and lose touch with the "*irreducible facts of reality*" which are not forthcoming and are unambiguous in their exponentially growing complexity. At the same time, it must guard against inducing rationalization where "one must destroy the (Vietnamese) village in order to save it." There is where loving (transcendence) must distinguish itself from apocalypse.

We can embark on the lifeboat innocently and full of love.

We can embark on the lifeboat passionately battling for its safe, if short journey.

We can embark on the lifeboat passionately creating beautiful poems.

We can embark on the lifeboat determined to understand its mechanisms and improve its performance.

We can embark on the lifeboat cynically expecting the fatal demise of it.

We can embark on the lifeboat unendingly transforming it and in so doing transforming its cargo.

This last is the true altruism of love and the subject matter for it (boat and cargo) is drenched in blood and light (matter and its metamorphosis).